World Food Supply

This is a volume in the Arno Press collection

World Food Supply

See last pages of this volume for a complete list of titles.

FOOD FOR DEVELOPING COUNTRIES IN ASIA AND NORTH AFRICA

(A SOCIO-ECONOMIC APPROACH)

V. G. RASTYANNIKOV

ARNO PRESS

A New York Times Company

New York — 1976

Editorial Supervision: MARIE STARECK

Reprint Edition 1976 by Arno Press Inc.

Reprinted from a copy in
The Princeton University Library

WORLD FOOD SUPPLY
ISBN for complete set: 0-405-07766-1
See last pages of this volume for titles.

Manufactured in the United States of America

Library of Congress Cataloging in Publication Data

Rastiannikov, Viktor Georgievich.
Food for developing countries in Asia and North Africa.

(World food supply)
Reprint of the 1969 ed. published by "Nauka" Pub. House, Moscow.
Translation of Prodovol'stvennaia problema v razvivaiushchikhsia stranakh Azii i Severnoi Afriki.
1. Food supply--Asia. 2. Food supply--Africa, North. I. Title. II. Series.
HD9016.A2R313 1976 338.1'9'5 75-26311
ISBN 0-405-07789-0

USSR ACADEMY OF SCIENCES
INSTITUTE OF THE PEOPLES OF ASIA

V. G. RASTYANNIKOV

FOOD FOR DEVELOPING COUNTRIES IN ASIA AND NORTH AFRICA

(A SOCIO-ECONOMIC APPROACH)

„NAUKA" PUBLISHING HOUSE
Central Department of Oriental Literature
MOSCOW
1969

Translated from the Russian
by GEORGE S. WATTS

The book is concerned with the main disproportion in the food balance of the developing countries—the ever-widening gap between the level of food demand and the actual potential to meet it from national production. In this connection, the factors behind the current food requirements are subjected to thorough analysis. Particular attention is attached to the changes in the population structure of the developing nations. Also discussed in the book are the possible ways of overcoming the dependence of the developing nations on the food supplies from the highly developed capitalist countries.

Виктор Растянников

ПРОДОВОЛЬСТВЕННАЯ ПРОБЛЕМА В РАЗВИВАЮЩИХСЯ СТРАНАХ АЗИИ И СЕВЕРНОЙ АФРИКИ

(на английском языке)

Printed in the USSR

CONTENTS

From the Author 4

The Main Disproportion 6

Agriculture — Its Technological Basis and Productivity. 14

The Marketed Food Resources: Trends in Demand and Supply . . . 20

The Growth in Food Imports 46

The Market and Reproduction in Grain Farming. The Influence of International Factors 53

The Market and Reproduction in Grain Farming. The Influence of National Factors 80

Reorganizing Agriculture: A Strategic Approach 9

FROM THE AUTHOR

The food problem is one of the most pressing issues facing a large number of developing countries at the present stage of their socio-economic development. The shortage of food has been especially sharply felt by these countries in the past decade, at times taking on catastrophic proportions. A real food crisis flared up in the mid-sixties in vast areas of the Asian continent populated by hundreds of millions of people. A number of countries proclaimed the problem of food supplies „National Problem No. 1".

Actually, the tasks that must be resolved are truly colossal. And they are determined not only by the difficulties of the present day. According to the FAO's Third World Food Survey (1963), in order to considerably raise the norms of food consumption in the developing countries, the food supply by the year 2000, bearing in mind the growth of the population, must be 4 times the present food resources in these countries. The emerging Asian countries alone must have grain stores of about 800 million tons, whereas today they produce only 170-175 million tons.

The enormity of this task requires a bold economic policy, especially if we take into account the present food crisis. In numerous developing countries the organizations responsible for the national economic policy are coming to realize more and more clearly that measures to liquidate this crisis and boost food production should be given priority within the framework of today's national economic policy. In a number of instances the reappraisal of the significance of the food problem entails a revision of established viewpoints on such questions as the role of agriculture in developing national economy as a whole, the relative amounts of national investments to be allocated to different sectors of the national economy, efficiency and goals of agrarian reforms, etc.

All this means that the food problem is by no means a "pure" case of production. It has many political and socio-economic aspects to it. The main contradictions in the economic growth of the developing countries are most graphically manifested in the critical state of the productive forces of agriculture. Taking advantage of the harsh food difficulties, the developed capitalist countries (first of all, U.S.A.) are stepping up their offensive on the countries of the "third world" in order to preserve them as an object of exploitation in the world system of capitalist economy and to retard the revolutionary process taking place there. Therefore, it is only natural that in the emerging countries there is an ever-increasing dissociation of class forces in respect to the ways of resolving the food problem; there are more clashes between the various classes in the sphere of political and social life. Hence, the question of food supply tends to acquire the dimensions of an important catalyst in the entire class struggle over the choice of ways these countries should take in their socio-economic development.

It is obvious that the food problem is a complex one, whose nucleus is made up of numerous factors: economic and social, natural and demographic. In the given work, the author made an attempt to discuss, to the best of his ability, only some of the aspects of the problem—those which seem most important and most interesting, concentrating on the developing countries of Southeast Asia and the Far East (countries east of Iran), an area suffering from the severest food shortage, and also on some countries of the Near East region (countries of West Asia and the United Arab Republic).

It goes without saying that the author alone is responsible for all the views expressed herein.

THE MAIN DISPROPORTION

In order to get a clear picture of the "bottlenecks" in the system of food supply in the emerging countries, it is first of all necessary to have an idea of how much food is consumed by their populations. In terms of political economy, one can say that in the developing countries which threw off the colonial yoke the mass of the national labour force, using Marx's words, is, in fact, in a "crippled state". The overwhelming majority of people in the developing countries still do not have enough to eat. If we take an index such as the minimum food fund necessary for the subsistence of an average person, it is obvious that it cannot be identical for all the countries. The size of this fund depends on the mean national intensity of the labour process (which is different for different countries), on climatic conditions and other factors. Though there are different approaches in determining the minimum food fund necessary for reproducing the labour power of a worker and his family (e. g., according to FAO, in various developing countries of Asia and North Africa, this per capita fund is in the vicinity of from 2,250 to 2,400 calories daily), at the given stage of socio-economic development of most countries in the regions in question, the average national per capita consumption of no less than 2,500 calories per day could be accepted as the minimally required fund (in the developed capitalist countries, according to FAO, the average per capita consumption standard is 2,900-3,200 calories, and in the USSR and East European socialist states, 3,010 calories per day).

However, according to FAO statistics, the average per capita food consumption in the developing Asian countries east of Iran taken together at the end of the 50's (1957-1959) was only 1,980 calories per day and in the early sixties (1962),

2,080 calories.[1] Nowadays even such a low rate of food consumption in many countries can be ensured only if backed by large and steadily growing food imports. But the 2,000-calory level was the level of consumption in the colonialist era. Thus, in the pre-war years (1934-1938) on the territory of former colonial India the daily per capita food supply was 1,950 calories; in 1960/61-1962/63 in the Indian Republic this index went up to 2,020 calories (although in the early fifties it was as low as 1,740 calories), and in Pakistan the level was 2,010 calories. In 1963/64 the per capita food supply in India dropped to 1,990 calories.

Only in a very few countries is there a noticeable and steady improvement in the per capita food supply. These include certain countries of the Near East, for instance, the UAR (2,360 calories in 1948/49-1950/51 and 2,890 calories in 1962/63), Turkey (2,510 calories and 3,110 calories respectively for the same years).[2]

On the whole, the national food consumption fund (irrespective of the sources it may come from) in the overwhelming majority of developing countries is, apparently, at least 20-25% less than the minimum subsistence level. And if we give due consideration to the qualitative structure of this fund, then its real value for reproducing the labour power will be even less (for instance, as a rule, it contains only half of the required quantity of proteins, moreover, the amount of animal protein being negligible; the per capita consumption of the latter in the early sixties in these countries amounted to only 6-16 gr per day). The acute protein deficiency which greatly increases the incidence of dystrophy, is a characteristic

---

[1] "FAO Commodity Review 1962. Agricultural Commodities Projections for 1970", Rome, 1962, pt. I, p. 26; "Monthly Bulletin of Agricultural Economics and Statistics", FAO, Rome, September 1967, p. 14. These figures are probably much exaggerated because they have been obtained by calculating the gross per capita food supply (entire produce less seeds). For example, if for 1958-1960 FAO gives 2,820 calories as the per capita food supply for Turkey and 2,530 for the United Arab Republic, the Economic Research Service of the U.S. Department of Agriculture which takes a fuller account of non-food uses of grain (seed and waste, industrial use, feed), cites other figures for approximately the same period (1959-1961), namely 2,600 calories for Turkey and 2,330 for the UAR (compare: "The State of Food and Agriculture 1964", FAO, Rome, 1964, pp. 228-229; "Food Balances for 30 Countries in Africa and West Asia", Washington, 1965).

[2] "The State of Food and Agriculture 1964", pp. 228-229; "The State of Food and Agriculture 1966", FAO, Rome, 1966, p. 217.

feature in the entire pattern of food consumption in the developing regions.

So far we have dealt with the average national per capita consumption level and the national food fund as a whole. However, the society's overall life sustaining requirements are made up, as is generally known, from the requirements of various strata and groups of the population. The largest part of society's needs in food is accounted for by the needs of the exploited population, most of whom are engaged in heavy physical labour. That is why the share of life sustaining requirements of the exploited section of the population exceeds their proportion in the total population. However, in distributing the food fund which a nation has on hand, because of the preservation of exploiting property and class inequality in society, an exactly opposite tendency is observed: the highest per capita consumption is in the top layers of society, while the lowest is in the lower brackets. In India, for instance, at the beginning of the 1960's the monthly per capita food expenditure of 38% of rural households was less than Rs. 11, while it exceeded Rs. 24 in case of 12% ot rural households (well-to-do and richer groups). Moreover, the per capita consumption of foodgrains (the staple product) in the higher groups was from 2 to 4 times greater than that in the lower groups.[3]

In the Philippines, according to the Bureau of Census and Statistics, in 1961 64.1% of the households with incomes of less than 1,500 pesos consumed less than 885 pesos' worth of food annually (including 17% of the households—only 475 pesos worth), while 5.8% of the households with incomes of 5,000 pesos' and more spent 2,628 pesos on food annually.[4] In other words, the volume of food consumption is directly proportional to the income grades. All this means that the mass of the people in the developing countries do not get the average national norm of the food supply and are chronically undernourished. According to a FAO's survey, "of the present world population, 10 to 15 per cent are undernourished and up to half suffer from some degree of hunger or malnutrition

[3] "The National Sample Survey. No. 101. Sixteenth Round: July 1960-August 1961. Tables with Notes on Consumer Expenditure (preliminary)", Delhi, 1965, pp. 15, 16, 19, 20.

[4] "Economic Research Journal", University of the East, Manila, vol. XI, 1964, No. 3, pp. 144, 147, 148.

or both".[5] That is why any shortage of food resources becomes a tragedy for the people who stand at the lower rungs of the social ladder.

This broad gap between the required and the existing norms of food consumption can be explained by a number of factors: the imperialist plundering of the developing countries (first of all, through a system of non-equivalent exchange); the presence of a large mass of unemployed and underemployed people in the sphere of production (with comparatively over-populated villages and cities); a preponderance of low productive economic structures, in which the overwhelming majority of the gainfully employed population is engaged.

The present food demand of the developing countries has grown considerably since the end of World War II. This growth stems directly from the natural increase of the population progressing at an unprecedented pace (which is often described as a "population explosion"). For instance, in the period from 1950 to 1960 the population in the developing countries of West Asia increased at an average 2.4% yearly, in the developing countries of Asia east of Iran—2.1%.[6] (For the sake of comparison, the population of the whole of Asia grew at a rate of 1.0% a year in the decade of 1920-1930, 1.3% in 1930-1940, 1.4% in 1940-1950). And beginning with the sixties this rate has stepped up still further. In India which has the second largest population in the world the growth of the population moved from 1.9% a year in 1951-1961 to 2.4% in 1961-1966; in Pakistan in the first half of the sixties the population increase was 2.7% per annum (against 2.2% in 1951-1961). In the Philippines the population increase in the 1960's was 3.4% annually. In the UAR the annual increase of the population in the first half of the sixties fluctuated between 2.6-3.0% yearly. And the forecasts for the expected population increases for different regions of the world in the next few decades are more than impressive. According to the UN estimate, only in the countries of the ECAFE region, with a rate of population increase equivalent to that of the late fifties and early sixties, the population in 1980 can swell to 2,454 million, and in 1989 it can exceed 3,000 million (1960—1,605 million). Throughout the whole of Asia (1960—1,652 million) the population will pass

---

[5] "The State of Food and Agriculture 1965", FAO, Rome, 1965, p. 6.
[6] "World Economic Survey 1963", UN, New York, 1964, vol. I, p. 20.

the 3,000-million mark several years earlier. In other words, in less than three decades the population of Asia can almost double and be equal to the entire population of the planet in 1960. [7]

As for individual countries, India's population, at the present rate of growth (2.4%), could rise to 689 million by 1980 (as compared to 439 million in 1961). According to the Central Board of Family Planning, Government of India, in 1995 the country's population will surpass the 1,000-million mark. Pakistan's population in 1980, at the present rate of growth (2.7%), will increase to 153 million (as against 94 million in 1961), the population in the Philippines will rise to 55 million (as against 27 million in 1960). It goes without saying that such a rapid population growth automatically means a swiftly growing demand for food.

However, the growing population is not the only reason for the increase of current food demands. The dynamics of these requirements also depends on the country's economic growth rates. The higher these rates, the bigger the per capita national income. In the United Arab Republic, for example, due to the major social changes and the acceleration, on this basis, of economic growth rates, the per capita national income rose by 61% (in stable prices) from 1952/53 to 1964/65. On the contrary, in a country like India it increased only 27% from 1948/49 to 1964/65, and in Pakistan the increase was only 12% from 1949/50 to 1964/65, [8] although the population in both countries grew at a slower pace than in the United Arab Republic. A definite part of the increase in the national income accrues to the working sections of the population, great in number and proportion, who are the ones suffering from chronic undernourishment. Under such circumstances, an increase in the consumption fund of the national income must proceed primarily along the lines of increasing food consumption, because food is not only the

---

[7] "Report of the Asian Population Conference and Selected Papers (1963)", UN, New York, 1964, pp. 67, 85, 86; "Bulletin on Food Statistics, February 1966", Delhi, p. 5; "Government of Pakistan. Outline of the Third Five-Year Plan (1965-1970)", S. 1., 1964, p. 10; "Statistical Handbook of the United Arab Republic 1952-1965", Cairo, 1966, p. 8.

[8] "Statistical Handbook of the United Arab Republic 1952-1965", p. 236; "Estimates of National Income 1964-65", New Delhi, 1966, p. 1; "Government of Pakistan. Outline of the Third Five-Year Plan (1965-1970)", p. 1.

first item of consumer demand but also the biggest one in volume.

Thus, the current food requirements of the population in the developing countries, reflected in effective demand, grow at a faster pace than the increase in population, and they will be all the greater, the higher the growth rates of the national income, on the one hand, and the bigger the share of the national income redistributed in favour of the working sections of the population, on the other. In the United Arab Republic food requirements of this kind (that is, expressed in effective demand) rise more speedily than, for example, in India or Pakistan.

Indeed, while in the United Arab Republic the per capita food supply, according to FAO data, increased by more than 22% from the early 1950's to the early 1960's (largely as a result of bigger imports of foodstuffs), in India during this period it rose only by 15%, although the initial basis of growth in both countries differed substantially: 1,740 calories per capita daily in India against 2,360 in the UAR. In Pakistan, the figure remained practically stable.

Using UN statistics, the Economic Research Service of the U.S. Department of Agriculture carried out an interesting project aimed at determining the rate of food demand growth versus the rate of population growth in the developing countries. The results of this study are presented in Table 1 where for the sake of comparison also given are the indices for two more developed countries.

As the figures in Table 1 show, in the developing countries up to 30-40% of the total annual food demand increase is due to the demand increase caused by the growth in real per capita income (except for the countries having an extremely slow rate of economic development, those which entered the era of social transformations too late, for instance, Iran, Pakistan). [9] In the more developed countries this figure stands at 70-80%. FAO calculations reveal that in 1965-1975 the rise

[9] In the first half of the 1960's, the rate of Pakistan's economic growth rose considerably and amounted to 4.8% per annum, according to the indices for the increase in the gross national product. Apparently, there has been a gradual change taking place in the structure of the food increase. For example, the per capita food supply grew up to 2,220 calories per day in 1963/64, as against about 2,000 calories in the 1950's and the beginning of the 1960's. This increase, to a considerable degree, was likely due to an increase in per capita income.

Table 1

A comparison of the rates of population increase and food demand for 1950-1960 (annual compound percentage change) *

Country	Rate of population increase	Annual increase in real per capita income	Annual increase in per capita food demand	Total annual food demand increase	% of annual demand increase accounted for by population growth
Philippines . .	3.2	1.7	1.28	4.48	71
Thailand . . .	3.2	2.4	1.73	4.93	65
India	2.0	1.7	1.36	3.36	60
Pakistan . . .	2.2	0.3	0.24	2.44	90
Iran	2.2	0.05	0.04	2.24	98
Turkey	2.9	3.2	1.57	4.47	65
UAR	2.4	2.5	1.62	4.02	60
Sudan	3.4	0.8	0.51	3.91	87
Tunisia	1.8	1.7	1.10	2.90	62
Japan	1.2	7.6	4.41	5.61	21
Greece	1.0	4.7	2.30	3.30	30

* "Changes in Agriculture in 26 Developing Nations 1948 to 1963". Washington, 1965, p. 4.

in per capita income could be credited with 16-32% of the entire increase in the food demand in all the developing regions of the world. The corresponding figures for Southeast Asia and the Far East might fluctuate from 18 to 35% and for the Near East countries, from 19 to 34%.

However, the long years of colonialist oppression which retained the backward forms of ownership in rural areas, ruining their productive forces, laid the extremely low initial basis from which agricultural development was to proceed in the period of independence. On the other hand, the agrarian reforms that were put into effect in many developing countries, due to a variety of reasons stemming mainly from class relations, have so far been unable to break drastically the old economic structure of agriculture. (The question of land reforms is dealt with briefly in the last section.) All this has a retarding effect on the growth rates of agricultural output.

A problem arises as to the extent to which the increasing food requirements of the developing countries can be satisfied by national production.

According to the ECAFE statistics, in the developing countries of Southeast Asia and the Far East the average annual rate of increase in food production for the period from 1953/54 to 1963/64 amounted to only 2.7%. In the sixties it was considerably lower and equalled only 2.2% per annum for the first five years (1960/61-1964/65).[10] Hence, the increase in food production during these years was lower than the natural increase in the population in the area (at least 2.4% annually). Moreover, the second five-year period of the sixties opened up with a drop in the absolute volume of food production: in 1965/66 it declined by 2%. The gap was particularly wide in India where the population swelled by more than 55 million from 1961 to 1966, while the grain output, far from increasing, even decreased by 1.9% annually.[11] The pre-war level of per capita food output in the developing countries of Southeast Asia and the Far East which had been regained, as ECAFE assumes, in 1960-1961, then again receded (in 1965/66 the per capita food production in the region as a whole was approximately 3% below the average pre-war level).[12] The tendency of dropping per capita food production increase takes place also in many countries of the Near East and North Africa (see Table 2).

Now, if we bear in mind that all these tendencies towards regress are observed at a time when the per capita food supply of the great masses of the population already does not exceed the semi-starvation level, the food problem boils down to the following: the developing nations (at least the majority of them), in spite of the tremendous potentials they may possess become less and less capable, at the existing rates of agricultural development, to supply their national food funds in the necessary quantities.

The incompatibility of the rate of growth of national food output with the rate of growth of the actual requirements of the emerging nations in food resources (even within the framework of a relatively low effective demand) is becoming more and more obvious, reflecting one of the most acute disproportions in the entire economic development of these countries.

---

[10] "Economic Survey of Asia and the Far East 1965", ECAFE, Bangkok, 1966, p. 214; "Economic Survey of Asia and the Far East 1966", ECAFE, Bangkok, 1967, p. 108.

[11] See: "Bulletin on Food Statistics. February 1966", pp. 1, 5; "Reserve Bank of India Bulletin", Bombay, June 1967, p. 747.

[12] "The Economic Survey of Asia and the Far East 1966", p. 108.

Table 2

The dynamics of crop output in 1948-1963 (annual compound percentage change) *

Country	1948-1963		1948-1955		1955-1963		
	Increase in total crop production	Increase in per capita production	Increase in total crop production	Increase in per capita production	Increase in total crop production	Increase in population	Increase in per capita production
Iran	3.6	1.4	3.8	1.6	3.3	2.5	0.8
Turkey . .	4.5	1.6	6.0	3.1	3.1	2.9	0.2
UAR . . .	2.0	—0.4	0.7	—1.7	2.8	2.5	0.3
Sudan . . .	8.0	4.4	10.2	6.8	5.8	2.8	2.3
Tunisia . .	1.6	—0.2	1.8	0.0	1.4	2.1	—0.7

* "Changes in Agriculture in 26 Developing Nations 1948 to 1963", p. 6.
Note: The statistics for the population growth for 1950-1960 were used to determine the per capita increase in crop production for 1948-1963; for the period from 1955 to 1963 the current rate of population growth (i. e. since 1960) was taken. Therefore, the drop in the per capita crop production increase is somewhat overestimated.

It is understandable that the actual backgrounds against which this disproportion manifests itself are different in different countries. In countries like the UAR where significant social transformations are taking place the reasons for such a disproportion lie in that the national output of food products, growing at a fairly fast pace, [13] cannot meet even the more rapidly increasing per capita demand for food. On the contrary, in countries like India or the Philippines the backward agriculture is not capable of producing enough food products to meet the actually stagnant (or slightly increasing) average per capita demand for food.

AGRICULTURE—ITS TECHNOLOGICAL BASIS AND PRODUCTIVITY

The very low rates of growth in agricultural production, in particular, the output of food crops, are caused by the fact that the agriculture of the developing countries regarded as a

[13] After 1952, for instance, grain production in the UAR grew 5% per annum (average for the period from 1952/54-1962/64).

branch of the economy, still lacks, as a rule, the necessary conditions for expanded reproduction. The dominating positions are occupied here by economically inefficient structures, which are characterized by extremely small-scale production units and very low labour productivity. Indeed, as the data of national surveys reveal, the large part of grain production is concentrated precisely in such structures (see Table 3).

Table 3

India, Pakistan: concentration of grain production in economically less effective structures in agriculture *

Country	Year of survey	Crop	Group of farms by size of holding (ha)	Proportion in area sown to the given crop (%)
India	1953-1954	Paddy Wheat	less than 6 less than 10	72.9 70.2
Pakistan	1960	Paddy Wheat	less than 5 less than 10	79.3 71.7

* "The National Sample Survey. No. 74. Eighth Round: July 1954-April 1955, Report on Land Holdings (5)", Delhi, 1963, pp. 363-364; "1960 Pakistan Census of Agriculture", vol. III, "All-Pakistan Report I", Karachi, 1964, pp. 15, 17.

It is this type of production that suffers most from the oppression by the still surviving pre-capitalist land property and trading and money-lending capital as well as from the exploitation by big banking and commercial bourgeoisie, by the state, etc. Hence, it is only natural that agriculture, which, as compared with other sectors of the national economy, produces the biggest part of the national income, can assign for productive accumulation only a negligible part of the new value it creates. In India, according to Reserve Bank of India, the savings of the rural household sector as a proportion of total agricultural income were only 2.3% in 1950/51-1960/61 and 2.2% in 1961/62-1962/63.[14] Moreover, over three-fifths of the rural households (62.6% in 1963) did not create any net savings whatsoever and were "eating up" the available resources (were being ruined); only in one-tenth of the rural households (10.2%) net savings amounted to the sizeable sum (by Indian standards) of Rs. 250 and more per

[14] "Reserve Bank of India Bulletin", March 1965, p. 327.

household.[15] As a result, some degree of progress in production that is observed in the emerging capitalistic structure counterpoises a tendency towards regress in vast sectors of other structures, such as pre-capitalistic, petty commodity production, etc. All these circumstances predetermined the extreme weakness of material facilities in agriculture. The use of modern agricultural equipment and fertilizers, in spite of the definite progress that took place in the post-war years, in the majority of the developing countries is negligible. For instance, in 1963/64 to 1965/66 the developing Asian countries east of Iran used on the average only 7 to 8 kg of chemical fertilizers (in terms of 100% of nutritional content) per 1 hectare of arable land (India 3.7 kg, Pakistan 5.0 kg, the Philippines 9.4 kg), and the Near East region — 6 to 8 kg (for the United Arab Republic, however, the figure was 109.8 kg already in 1962/63), while the African countries applied only 2 to 3 kg of fertilizers per 1 hectare of arable land.[16] In the first group of countries there was only 0.38 and in the second group 1.31 tractors per 1,000 hectares of arable land in 1964.[17] (The corresponding figures concerning individual countries are given in Table 4.)

Though many developing countries have ample water resources, irrigated farming which ensures stable harvests even on the old technical basis, is advancing very slowly. The share of irrigated farm fields in the general fund of cultivated land, in such countries as Thailand and the Philippines, amounts to less than 1/6, in India — 1/5 (in 1952/53 —17%), in Pakistan — about 2/5. It should be noted, however, that in some countries (for instance, India and Pakistan) the growing irrigation potential, due to a number of social reasons, is not fully utilized. Several countries face an extremely acute problem — vast areas of fertile land perish and become unfit for cultivation (e. g., in West Pakistan in the early 60's about 70% of all cultivated land was affected by waterlogging and saline processes; almost 40,000 hectares of land here perish from these processes every year).[18]

---

[15] "All-India Rural Household Survey", NCAER, New Delhi, 1965, vol. II, p. 96.

[16] "The State of Food and Agriculture 1965", p. 93; "The State of Food and Agriculture 1967", FAO, Rome, 1967, p. 46.

[17] "Monthly Bulletin of Agricultural Economics and Statistics", May 1966, p. 1.

[18] "Pakistan 1961/62", Karachi, 1962, p. 61.

Table 4

Changes in the technical basis of agriculture *

Country	Consumption of fertilizers, kg/ha of arable land **		Number of tractors per 1,000 ha of arable land	
	1948/49—1952/53	1962/63	1949/50	1961/62
Philippines . .	6.7	9.4	0.19	0.60
Thailand . . .	0.4	2.1	no data	0.50
India	0.6	3.4	0.05	0.21
Pakistan . . .	0.2	5.7	no data	0.15
Iran	no data	0.8	no data	0.36
Turkey	0.6	2.7	0.16	1.68
Sudan	0.7	2.9	0.02	no data
Tunisia	2.7	2.7	1.37	no data
UAR	47.1	109.8	no data	4.28
Japan	144.8	270.1	no data	234.4 ***
Holland . . .	124.6	518.5	no data	104.3 ****

* "Changes in Agriculture in 26 Developing Nations 1948 to 1963", pp. 47-48; "Production Yearbook 1964", FAO, Rome, 1965, pp. 3, 302.
** in terms of nutritional content — N; P_2O_5; K_2O.
*** including 232.8 garden tractors.
**** in 1963.

By and large, it can be asserted that the technological progress which is gradually taking place in the sphere of industrial production (at least in the more developed countries of the "third world") so far has practically not affected agriculture. On the whole, in many countries agriculture is continuing to represent what UN terminology calls the "traditional" sector of the national economy, where the entire system of production is based on the use of a colossal amount of manual labour with its extremely low average productivity. Even in the evolving capitalistic structure, a very large part of all capital employed is represented by variable capital, while its technical elements are comprised mostly of the means of production inherited from the feudal formation (draught livestock, simple implements, very restricted use of manures, etc.).

As a result, the output per person engaged in agriculture in the developing countries is very low, and in many of them practically insignificant. In several countries having the most backward agriculture with an especially low level of use of technical means (including those in the form of fixed assets), the output per person engaged in this sphere of the economy amounts to only $ 100-200 a year (see Table 5).

Table 5

Gross fixed capital formation in agriculture and agricultural output per agricultural worker *
(in U.S. dollars)

Item	Philippines	Thailand	India	Pakistan	Turkey	UAR	Japan	Greece
Gross fixed capital formation (1953-1961)	4	1	3	6	no data	19	47	29
Agricultural output (1960)	181	94	114	182	326	365	402	391

* "Changes in Agriculture in 26 Developing Nations 1948 to 1963" pp. 15, 64.

In the light of these statistics, it is possible to understand to what extent the food crisis is influenced negatively by the pitiful legacy left to the developing countries by imperialism — relatively high "over-population", primarily in the rural areas. In India by the late 1950's, according to estimates of H. Singh Sekhon, the extent of underemployment in the agricultural sector was equivalent to 25% of the total gainfully employed agricultural population. [19]

In other words, from the angle of production, a quarter of the population in the sphere of agriculture are "absolutely redundant" people in India. No matter how little food these "surplus" people consume, the responsibility of feeding them objectively devolves upon the productive part of the agricultural labour force which itself, due to the existing conditions, can produce (per person employed) only a minimum amount of food.

The average productivity of labour in the agriculture of the developing countries is, as a rule, several times lower than in industry. For instance, in Indian agriculture in 1960 every worker put out only 114 dollars' worth of produce, while as far

[19] "AICC Economic Review", New Delhi, 1960, September 7, pp. 9-10.

back as 1951 one factory worker engaged in large-scale industry accounted for 1,781 dollars' worth of output, moreover, the new value created amounted to 509 dollars. [20] Such a broad gap between the average level of labour productivity in agriculture and that in the urban large-scale capitalistic structure is explained to a certain degree by the socio-economic policy followed by the state (emphasis is placed on industrial development with relatively insufficient measures taken to raise the productive forces in agriculture).

Because of the limited use of modern means of production and due to traditional agro-technique, there is an extremely low output per unit of land in agriculture, especially in grain farming. For instance, in the early 60's the paddy yield in the main rice growing countries in Asia amounted to only 12-20 centners [21] per hectare (1 cent. of paddy = 0.66 cent. of cleaned rice), while wheat yields, as a rule, did not exceed 8-10 cent. per hectare (see Table 6). However, it should be noted that in the post-war years grain yields in many countries grew at rates amounting to only tenths and even hundredths of 1%. Moreover, most of the increase in yields (although not in all regions of this or that country) fell on farms of capitalistic type, making a greater use of modern technical means, rather than on smaller farms. Among the bigger countries, by the beginning of the 60's, India registered the biggest increase in yields. But even in this instance, the increase of yields in rice farming in the period from 1952/53 to 1961/62 (assuming that the yield for 1952/53-1954/55 = 100) amounted to only 1.63% per annum, while for wheat farming it was even less—0.94%. [22]

It was precisely the low productivity of land that reflected in its most concentrated form the crisis in the socio-economic structure of agriculture. This gave rise to a grave agricultural crisis (soil exhaustion, etc.) which requires (in the technical-economic aspect) tremendous capital investments, if it was to be overcome.

The agricultural production of various countries does not develop evenly. The reason for it should be sought in a multitude of socio-economic factors. Moreover, this uneven agri-

---

[20] "The National Sample Survey. No. 15. Report on Sample Survey of Manufacturing Industries 1951", Delhi, 1958, p. 27.

[21] The centner used here is equivalent to 100 kg.

[22] "Agricultural Development. Problems and Perspective. April 1965", Faridabad, 1965, p. 91.

cultural development within the framework of the "third world" countries is steadily intensifying. For example, due to the extensive use of technical means, especially fertilizers, combined with intensive irrigation, the UAR, by the level of its mean labour productivity and crop yields, has forged considerably ahead of practically all the developing countries of the two regions in question. This nation stands out prominently when compared to such countries as Iran, Pakistan, India, the Philippines (see Tables 4, 5, 6).

Table 6

Yields of main grain crops (100 kg per ha) *

Country	Wheat		Country	Rice (paddy)	
	1949-1953	1961-1963		1949-1953	1961-1963
India	6.7	8.4	Philippines	11.8	12.2
Pakistan	8.7	8.1	Thailand	13.1	14.3
Iran	9.0	8.6	India	11.3	14.8
Sudan	11.8	16,0	Pakistan	13.8	15.9
Tunisia	4.9	3.4	Iran	19.3	19.6
UAR	18.4	25.1	UAR	37.9	52.3
			Burma	16.0	17.3
U.S.A.	11.2	16.9	Japan	40.0	50.5
Holland	36.5	43.8	U.S.A.	25.6	39.5

* "Economic Bulletin for Asia and the Far East", ECAFE, Bangkok, June 1965, pp. 9-10; "Changes in Agriculture in 26 Developing Nations 1948 to 1963", p. 46.

THE MARKETED FOOD RESOURSES: TRENDS IN DEMAND AND SUPPLY

A definite part of the national food resources, before they reach the consumer, must inevitably pass through the sphere of commodity exchange. As a nation's economy develops and the social division of labour is extended, this part must increasingly grow. But it is this process that is lacking in many developing countries. The ever-widening gap between the volume of the national requirements in marketable food-stuffs and the actual supply of the marketed food produce available is another manifestation of the acute disproportion in the economic growth of the developing nations. It should be noted that this disproportion is felt quite painfully not only in the countries with a comparatively low per capita food supply, but even in those where this index is relatively high (and gradually increasing) (for example, Turkey and the United Arab Republic).

Several interconnected processes have given a strong impetus to the demand for marketed food-stuffs in the developing countries during the years of independence. In the post-war period, many of these countries witnessed a sharp increase in the rate of urbanization, due mainly to the migration of the rural population to cities.[23] As a result, there was even a greater gap between the urban and rural population (in a number of countries this process took on very great proportions) in favour of the former. For instance in Pakistan for the period from 1951 to 1961 the proportion of the urban population rose from 10.4 to 13.1%, moreover, the latter grew by more than 56% (cities accounted for 25% of the total increase in the population).[24] In 1965 already about 15% of the country's population lived in urban areas. There was even a faster rate of growth in the city population in the UAR. In the period from 1947 to 1960, during which the country's population swelled by 37%, the urban population grew by 66%, the latter's share of the total population rising from 30.1% to 36.6%. At the same time, the cities accounted for more than 54% of the total population increase in the UAR.[25] In other

---

[23] Describing the nature and consequences of this migration, the 1958 ECAFE Economic Survey notes that since the end of World War II the population in most of the cities in the countries of the ECAFE region grew very rapidly, mainly due to migration from rural areas. "This migration has not by any means been caused only by industrialization. Social unrest, partition (of families — *V. R.*), rural population pressure (on the sources of means of livelihood — *V. R.*), and in some cases conditions of insecurity in rural areas are also contributing factors. Actually the recent rate of urbanization has usually been faster than the recent rate of industrialization." As a result, the migrations "swell the population in the overcrowded slum areas and often overflow into or create new shanty towns". "The net result of all these factors is a general lowering of standards and a reduction of facilities per capita for the entire city population". ("Economic Survey of Asia and the Far East 1958", Bangkok, 1959, p. 155). The problem, however, is that a reduction in the per capita food consumption goes parallel with an increased need in marketable food-stuffs, because the urban population cannot but use only marketable food.

[24] "Economic Survey of Pakistan 1963/64", Rawalpindi, "Statistical Section", p. 3.

[25] "Statistical Handbook of the United Arab Republic 1952-1965", p. 3.

words, the overwhelming majority of the newly forming population was concentrated in the urban localities.

In this respect, the picture is really impressive if one compares the figures for the population growth for certain countries vis-à-vis the population growth in their biggest cities. [26]

City	Period	Annual rate of growth of city population (%)	Country	Annual rate of growth of total population (1950-1960) (%)
Bombay . .	1951-1960	3.3	India	1.9
Delhi . . .	1951-1960	4.3	»	1.9
Djakarta . .	1955-1959	10.8	Indonesia	2.0
Teheran . .	1956-1960	5.0	Iran	2.2
Karachi . .	1951-1961	4.2	Pakistan	2.1
Manila . .	1948-1957	4.5	Philippines	3.1
Bangkok . .	1947-1960	6.9	Thailand	3.2

These figures show quite graphically the predominant growth of the urban population.

Along with the growth of the manufacturing industry, further specialization of agriculture, in particular, both in certain regions as well as in certain groups of farms giving priority to this or that technical crop for the market, gave a new impetus to the growing demand for marketed food products in the rural areas. At the same time there was a stepped-up stratification of the rural population into classes typical of the capitalist society. In particular, this led to a swelling of those sections that tended to purchase goods of prime necessity (including food-stuffs) exclusively or primarily on the market. In other words, the post-war upheavals in the economy of the developing countries, connected with the further growth in the social division of labour, the expansion of commodity and money relations, on the one hand, and the

---

[26] "Report of the Asian Population Conference (1963)", p. 82. Urbanization evidently will proceed at accelerated rates in developing countries in future as well. According to UN estimates, the share of the urban population in the developing nations the world over might reach 30.3% by 1985. Undoubtedly, growing urbanization is a long-term factor which makes for a consistent and stable rise in the demand for marketable foodstuffs.

expropriation of the producers, on the other hand, changed the structure of the population in such a way that the share of the population consuming marketed food products grew considerably; moreover, it grew at a pace greatly exceeding the overall rate of population growth. Furthermore, the greater part in the increase of the demand which stems from the increasing per capita income can be satisfied only through the sphere of the market (since today the growth of the gross national product in the developing countries, in the main, is due to the development of the commodity sector of production). All this means that the problem of supplying the population in the developing countries with food depends, to an ever-increasing degree, directly on the realization of farm products through the market.

The main staple in the diet of the population in the developing countries in Asia and North Africa (2/3 = 3/4 of all the calories supplied and more) is grain. Cereals account for the greater part of the in crease in per capita food consumption (in those developing countries where such an increase is registered at all). [27] And it is precisely the scale of and conditions for market realization, first of all, of grain products, that determine to what extent this or that country is able to provide the marketable component of the national food fund.

Grain farming in the developing countries is noted for its extremely low output of marketed products. It is noteworthy that in many countries the level of marketed grain has remained more or less stable for several decades, and in some countries, as we shall see later, it has even dropped. In India, for example, according to the data provided by the former Imperial Council of Agricultural Research, only 30% of the gross grain output was sold on the market in the 1930's and first half of the 1940's. But by the end of the forties as little as 27.5% of the total rice produced, which is the most important grain crop in the country, reached the market. [28] In the mid-fifties the total marketed grain was estimated by the Ministry of Food and Agriculture to be 32%, with 31.4% of the gross rice harvest and 37% of the gross wheat produ-

---

[27] For instance, the share of grains in the overall increase in per capita food supply that took place since the beginning of the 1950's, according to FAO statistics, was: 62% for India, 57% for Turkey and 74% for the UAR ("Production Yearbook 1964", p. 262).

[28] "Agricultural Marketing in India. Report on the Marketing of Rice in India", Delhi, 1954, pp. 418-419.

ce finding their way to the market.[29] An estimate by the Ford Foundation Commission for the period from 1949/50 to 1957/58 determined that the marketed grain in India was less than 25% of its total production.[30] The Reserve Bank of India, in a study of the economy of agriculture in 1961-1962, established that the rural areas that year supplied the market with 23.7% of the total rice harvest and 27.6% of the total wheat harvest.[31]

At the end of the fifties in Pakistan grain farming realized through the market (according to a national estimate) only 25% of its output. An estimate by certain economists places the consumption *in natura* of food crops at that time at 80% and correspondingly the marketed share was only 20%.[32] These figures are fully confirmed by the appraisal of FAO which for the period of 1947-1958 established the average level of marketed grain output in Pakistan to be approximately 20% for rice and 25% for wheat.[33]

According to a survey conducted in the Philippines in 1956 only 31.2% of the gross rice harvest — the main food crop in the country — was sold on the market.[34]

This picture hardly changed in the sixties. According to a study made in 1962-1963, 47.1% of the rice grown in the country entered so-called commercial channels.[35] Of this figure 10.1% was used as payment in kind to the farm workers and farmers who helped their neighbours take in the harvest, i. e., it entered the consumers' fund directly. Approximately 7.5% of the rice (as can be judged from the data of other surveys of the economics of rice farming)[36] was appropriated in kind by the money-lenders; apparently, from 1/2

---

[29] "National Food Reserve Policies in Underdeveloped Countries", FAO, Rome, 1958, p. 27; "Indian Agriculture in Brief", Delhi, 1965, p. 163; "Agricultural Marketing in India. Report on the Marketing of Wheat in India", Nagpur, 1963, p. 21.

[30] "Report on India's Food Crisis and Steps to Meet It", Delhi, 1959, p. 98.

[31] Calculated from: "All-India Rural Debt and Investment Survey 1961/62. Tables Relating to Gross Farm and Non-Farm Receipts" (mimeo), Bombay, S. a., pp. 6, 9, 23, 24.

[32] "Enterprise", Karachi, 1958, March 1, p. 10; Mahbubul Haq, Deficit Financing in Pakistan 1951-1960, Karachi, 1961, p. 68.

[33] "National Food Reserve Policies in Underdeveloped Countries", p. 37.

[34] "Economic Research Journal", Manila, vol. III, 1957, No. 4, p. 211.

[35] *Ibid.*, vol. XII, 1965, No. 1, p. 14.

[36] *Ibid.*, vol. XII, 1965, No. 3, p. 204.

to $^2/_3$ of these appropriations (4-5% of all the rice) was subsequently directed to the growers in the form of loans in kind and did not enter the market. Therefore, it seems, only 32% of the rice output ultimately reached the market for sale.

On the whole, it may be considered that on the threshold of the 1950-1960's in the developing Southeast Asian countries (excluding certain countries with a mono-grain crop agriculture) the level of marketability in grain farming amounted to approximately 25-30%. (For the sake of comparison, in the developed capitalist countries, such as the U.S.A., Canada, Australia from 80 to 90% of the grain output is directed towards the market). The low output of marketed grain in the developing countries reflects the predominance of poor production structures in farming economics.

Hence, it is only natural that petty production plays an extremely great role in the developing countries in supplying the market with food products for sale. In the quantitative respect a petty producer who cultivates land wholly or mainly with his own labour with the produce mainly consumed by his household and who supplies to the market only a smaller part of his output is the chief type of rural commodity producer. For instance, according to the 15th Round of National Sample Survey conducted in 1959-1960, in India, out of the total number of households with cultivated land which marketed their produce (79.6% of all the households having cultivated land), 89% came under the above-mentioned group of commodity producers. Along with the farms based entirely or mainly on the use of hired labour, which also consume the greater part of their own produce (and sell less), the proportion of commodity producers consuming the bulk of their output rises to 97%. Only 3% of the commodity producing farms sold the greater part of their output on the market. Meanwhile 20.4% of the total number of households consumed all the produce they grew.

Therefore, in view of the predominance of petty production structures in agriculture with the resulting extremely low efficiency of production, underemployment of the labour force, etc., the chief motive behind production in the vast mass of holdings, in contrast to countries with highly developed agriculture, remains providing food, first of all, for the producer himself and for his family.[37] That is why it

---

[37] The Philippino economist, D. A. Maulit, noted in this respect: "Rice farms (in the Philippines — *V. R.*) regardless of the tenure of the

is clear that with extremely slow rates of increase of grain harvests, the amount of grain that is placed on the market can also grow very slowly. This alone widens even greater the gap between the total national demand for marketed grain and the real possibilities for meeting this demand from the output of the nation itself.

However, the situation becomes even graver since the economic basis for the formation of marketed grain fund in the developing countries has never been stable. The movement of the marketed mass of grain (its growth or reduction) is determined by a broad complex of socio-economic reasons, moreover,—and this is very important—the volume of grain arrivals on the national market is not in direct dependence on the movement of grain production. Under specific conditions of developing economies, what is that which determines the amount of marketed grain? At the present stage, what general tendency characterizes the volume of market arrivals of grain?

In many developing countries the final volume of grain that reaches the market still depends to a very great extent on the system of forced commercialization of petty farms, subsistent in essence but, as shown already, accounting for a very large share of the national grain output. One of the largest sources of supplying the market with grain through the system of forced commercialization in a number of countries remains petty, tenant farming, bending under the burden of very high rates of rent, in the majority of cases paid out in kind (share cropping, etc.). Even in such a country as India where in the period of independence, as a result of agrarian reforms, there has been noticeable progress made in the agrarian system, a large part of the marketed grain fund is formed in many regions by rental payments in kind. The following table, compiled on the basis of the 15th Round of National Sample Survey (1959-1960), as well as certain other statistical surveys [38] vividly demonstrates this point (see Table 7).

---

farm operator with the possible exception of the farms operated by managers, are basically subsistence farms... It is for this reason that the amount of palay produced for the market is only 31% of the total production of the country". (D. A. Maulit, Statistics on the Disposition of Philippine Palay Production, *vide*: "Economic Research Journal", vol. III, 1957, No. 4, p. 206).

[38] See: "Indian Agriculture in Brief", Delhi, 1965, p. 163; "Agricultural Situation in India", Delhi, March 1962, p. 1212; "Survey on Distribution of Agricultural Produce", Trivandrum, 1961, p. 15.

Table 7

The role of rent in kind in the formation of marketed grain fund in certain states of India (1959/60)

State	Crop	Marketed produce as % of total output and date of estimate	Rent in kind as % of total output	Rent in kind as % of marketed produce, under the assumption that	
				the total rent in kind is realized on the market	²/₃ of the rent in kind is realized on the market *
Kerala	rice	23 (1959)	16.9	73.5	49.1
Madras	rice	25 (1956/57)	14.5	58.1	38.7
Mysore	rice	53.3 (1956/57)	17.7	33.2	22.1
Punjab	rice	64 (1956/57)	18.7	29.2	19.5
West Bengal	rice	32 (1956/57)	6.3	19.8	13.1
Andhra Pradesh	rice	50 (1956/57)	6.2	12.3	8.0
Punjab	wheat	31.6 (1960/61)	15.7	49.7	33.2

* Here it is assumed that one-third of the rent received by the landowners is consumed by their families.

These figures give a picture of the situation in this or that state on the whole. However, it should be noted that the significance of rent in kind for the formation of the marketed grain fund in a number of districts, specializing in the commercial production of grain and taking upon themselves a significant part of the burden of marketed arrivals of grain in the state, becomes even greater: large pre-capitalist land ownership here is quite often firmly entrenched and the rents for leased-in lands are especially high. For instance, such are the fertile regions of the river deltas along the east coast of India (Andhra Pradesh and Madras states), certain districts of West Bengal and others (see Table 8).

A similar picture is seen in other countries. For instance, in the Philippines, according to the 1956 survey, the bulk of marketed rice—58%—came from the semi-feudal tenants' farms, moreover, the rent in kind (including payments of debts to landlords) swallowed 18.1% of the total rice harvest in the country. [39] In 1962-63 when the percentage of

[39] "Economic Research Journal", vol. III, 1957, No. 4, pp. 211-212.

Table 8

Proportion of land rent in kind to the gross and marketed product in certain regions of India *

Region	Year of survey	Rent in kind as % of gross output of the region	Rent in kind as % of marketed product of the region
Madras, Tanjore District . .	1959/60	24.9	41.0
Andhra Pradesh, Krishna District	1959/60	24.1	33.9
West Bengal, Burdwan District	1957/59	18.3	31.9

* According to data: "Rural Credit Follow-up Survey 1957-58", Reserve Bank of India, Bombay, 1961, pp. 14, 18, 20; "Rural Credit Follow-up Survey 1959/60", Bombay, 1962, pp. 13, 16, 17, 274.

Note: The data concern the regions where rice is the main (though not the only) commercial crop. Rent in kind almost entirely pertains to grain farming. That is why the percentage of rent payments in grain within the marketed grain fund is, apparently, higher than the percentage shown in the Table.

land rent in kind reached 17.9% of the total rice harvest, [40] it accounted for approximately 56% of the entire marketed rice placed on the national market for sale.

These rent payments, in the main (or in other cases almost entirely, like in West Pakistan or in the Philippines), were concentrated in the hands of the class of big land owners. And it is precisely this class which is one of the chief holders of the marketable grain. [41]

It is evident, therefore, that in those cases when the rural areas are still in the grip of pre-capitalist forms of land ownership, the volume of grain that is placed for sale on the market is determined not by economic laws governing the development of production for the market, but by the ability of big semi-feudal land owners (taking advantage of their land monopoly) to squeeze out of the producers the utmost amount of rent, which is then transformed into mar-

[40] *Ibid.*, vol. XII, 1965, No. 1, p. 14.

[41] In West Pakistan where prior to the reform of 1959 the tenanted area (primarily share-cropping) enveloped over 3/5 of the cultivated land, the land owning class appropriated up to 25% (primarily as rent in kind) of the gross farm product (see: "The First Five-Year plan 1955-1960 (draft)", Karachi, 1956, vol. I, p. 165). Actually, the greater part of the marketed grain was concentrated in the hands of this class.

keted produce. And it is only natural that the stronger the position of big pre-capitalist land ownership in the country (or region of the country), the greater the role of rent in the formation of marketed grain that reaches the home market.

The semi-feudal system of agriculture, based on petty tenanted farming is noted for its extremely backward technical basis of production, low labour productivity and very small output per unit of land area. And this sets rigid limits on the possibilities of increasing the marketable volume of grain from the sphere of semi-feudal system of farming. This system of production relations in "normal" conditions (i. e., if the rent absorbs only the surplus product of the tenant) will ensure a relatively stable rather than growing volume of marketed grain for the internal market. Only by taking over part of the necessary product of the producer and changing it to commodity which in itself violates the reproduction process on tenanted farms (first of all, by reducing the producer's consumption fund) is the given system able, to a certain degree, to increase the supply of marketed grain in the existing conditions of production. And it was precisely this that was typical of many developing countries in the past decades.

A significant role in forced commercialization of petty production is played by the exploitation of the peasant by trading intermediaries and money-lenders. Money-lending in this case develops in several, though inter-related directions. In itself the pressure brought to bear on the borrower—the petty producer—due to the outrageously high rate of interest compells him to sell a great deal. In this, the volume of "surplus" grain sold by the producer, on the whole, is directly proportional to the interest rate which he must pay out. And since in grain farming in many developing countries (in the Philippines, for instance) the interest is often taken in kind, large quantities of the product are concentrated in the hands of the lender. Later he himself sells it on the market. But this does not change the main thing—the formation of the marketed product, in both cases, stems from the fact that the producer is forced to give up his grain.

In a number of countries (for instance, in India and Pakistan) there is a very widespread method of raising the volume of marketed grain by the money-lender's exploitation, whereby the creditor forces the debtor-producer to sell him his products at prices lower than those on the market (in such cases the nominal interest rate not infrequently is utterly disregarded—it exists in the difference of the prices).[42] In order to cover the interest charge, the producer must sell an additional part of his produce. Moreover, it is understandable that this part is the greater, the wider the difference between the market price, on the one hand, and the price at which the borrower sells his produce, on the other. This system of compelling the borrower to sell his produce (or direct alienation in kind) in favour of the lender, whatever form it may take, leads to a situation in agriculture in which the producers must relinquish a considerable part of the values they produce. As Soviet and foreign studies have repeatedly shown, under the influence of such exploitation by money-lenders the great mass of petty producers in the developing countries sell their grain at prices which actually do not cover the cost of production. In the long run this gives the same results as the burden of extremely high rent, that is interfering with the process of reproduction on the farm and cutting down the producer's consumption fund.

Money-lenders' capital forms one of the mighty pillars on which the entire system of mobilizing marketed grain is based in many developing countries. A study of rice farming in the Philippines in the late fifties revealed that the petty land holders, even under the assumption that they sold their rice produce (or gave it directly to their creditors) at prices which covered the cost of production, had to give up approximately 17% of their produce as payment in kind for interest only on current debts. However, since the average interest rate was 98%, and they paid out only 46%, it is apparent that if they fully paid back only the interest on their current debts, the latter would take up at least a third of all the rice that was both sold and paid out by these petty land holders. Even in tenant farming, burdened by unusually high rent in kind, it was still possible to increase the marketed "surpluses" of rice by money-lender's bondage. All the rice that came from the tenants' farms (44.4% of the total harvest) could be classified in the following manner:

---

[42] In Pakistan, according to the Credit Enquiry Commission, "the borrower is bound to sell his produce only to the lenders who purchase it at prices well below the market quotations. In effect, the profit obtained on the trading transactions more than makes up for the absence of a formal interest charge". ("Credit Inquiry Commission Report", Karachi, 1960, p. 10).

71.8% for rent, 9.5% which the tenants sold freely and 18.7% for payment in kind for the lenders' credit and interest. In view of the fact that the interest rate paid out by the tenants really amounted to 61%, the lenders' share was therefore divided up in the following manner: 11.6% for the outstanding debt (subsequently renewed), 7.1% for interest—clear profit for the money-lender, which ultimately takes on a marketable form, just like the rent in kind. (It is evident that the pure share of marketed "surpluses" in tenant farming is approximately 2/5 of the total grain harvest). [43]

Even though in India transactions in kind between the producer and the money-lender are apparently not so widespread as in the Philippines, nevertheless, in a number of states the mass of produce that the lenders squeeze out of the producer in repaying debts and interest occupies a formidable place in the overall volume of grain that is marketed. On the basis of the data from the 15th Round of National Sample Survey (1959-1960), it is possible to determine the money-lenders' share in the gross and marketed rice produce that they took in kind from the harvest of the producer:

State	Repayments in kind to money-lenders as % of gross produce	Repayments in kind to money-lenders as % of marketed produce under the assumption that	
		the total mass of repayments is sold on the market	1/2 of the mass of repayments is sold on the market *
Orissa	10.5	33.8	16.9
Madras	5.7	22.9	11.4
Andhra Pradesh	8.4	16.9	8.4
West Bengal . . .	4.3	13.4	6.7

* It is assumed that 1/2 of the debt repaid in kind is again renewed as debt by the producer.

[43] According to the "Economic Research Journal", vol. VI, 1959, No. 2, pp. 76, 78, 79; *Ibid.*, vol. III, 1957, No. 4, p. 211. In the light of these data, it is interesting to make the following parallel comparison: according to a survey conducted in the early thirties, in the Attock District of Pakistan where the peasants owned a considerable part of the land, 50.1% of all the wheat that came from the peasant holdings went for payment in kind for usurious credit and interest; 43.3% was used up for rent in kind and only 6.6% of the grain was sold freely by the producer. On the whole, 42.8% of the total grain harvest was either appropriated by compulsion or sold freely on the market. The marketed "surpluses" were,

Bearing in mind that in comparatively large areas of the country (first of all, those which have a relatively high level of commodity and money relations) there is a predominance of cash transactions between the money-lender and the producer (which means that the producer sells his produce himself and then repays the lender his debt from the money he receives on the market), one can see at once that the money-lender's exploitation of the agricultural producer has considerable influence on the amount of marketed grain.

The commercialization level of grain farming is largely affected (and in a number of countries to an ever-increasing extent) by the burden of taxes.

As a result, in the total amount of marketed grain the share that was formed by compelling the producer to relinquish part of his produce was unusually high. According to several surveys, in the 1950's in India up to one-half of the grain realized on the market was sold under compulsion. The other half came from farms having a commercial surplus [44] (moreover, national estimates usually do not take into account the grain that is paid out as rent in kind). Commenting, in particular, the figures submitted by Dharm Narain of the Institute of Economic Growth for 1950/51, the distinguished Indian economist, V. K. R. V. Rao wrote: "...only half of the marketed surplus is what may be called a commercial surplus, while the other half is what may be called a *distress surplus* or a surplus that is necessitated by the basic cash obligations of the agricultural producer rather than by the excess of his production over his desired consumption".[45] The situation has hardly changed since then. According to th

---

apparently, less, since part of the debts paid back to the lenders by the peasants were subsequently renewed in kind. With an interest charge of only 25% on credit in kind, the marketed "surpluses" in the district would be about 26% of the total harvest (for details see: V. G. Rastyannikov, On the Nature of Commercial Grain Farming in the Punjab, *vide*: "Brief Notes of the Institute of Oriental Studies", Moscow, No. 23, 1957). In this case it is interesting to note that the general results of the effect of pre-capitalist forms of land ownership and capital on the formation of the marketed grain mass coincide, in spite of all the differences of national peculiarities and historical eras to which both examples are related.

[44] D. Narain, Distribution of the Marketed Surplus of Agricultural Produce by Size-Level of Holdings in India. 1950/51, Bombay, 1961, pp. VI, 35; "The Economic Weekly", Bombay, 1961, January 21, p. 87.

[45] V.K.R.V. Rao, Foreword, *vide*: D. Narain, *Op. cit.*, p. VI.

data of a survey conducted by the Reserve Bank of India in 1961-1962, the distribution of marketed rice supplied by different groups of cultivators looks like this: [46]

Asset group of cultivators' households	Group as % of total cultivators' households	Value of gross produce per cultivator's household		Rice produce marketed by the group as % of total marketed rice produce		
		rupees	U.S. dollars			
Rs. Less than 5,000	64.3	from 208 to 619	from 44 to 130	19.8	38.6	59.6
Rs. 5,000-10,000	18.6	1,007	211	18.8		
Rs. 10,000-20,000	10.7	1,695	356	21.0		
Rs. 20,000 and more	6.4	3,930	825	40.4		
Total	100.0	901	189	100.0		

If we take into consideration that not only the first two groups of households, but a certain part of the third group (with assets from Rs. 10,000 to 20,000), practically had no normal commercial surplus of grain, one can come to the conclusion that 40-50% of the marketed rice in India in the early 1960's was formed, in the main, by compulsory sale of grain by petty producers.

In the Philippines, as can be judged from the material of national surveys, at least 70% of the marketed rice fund was formed by payments for rent and usurious debts.

For instance, a survey conducted in one of the regions of Central Luzon—the most important rice growing area in the country (in the second half of the fifties more than 37% of the national marketed rice came from Central Luzon), revealed the following pattern of distribution of the gross rice produce (average data for two years surveyed: 1955 and 1959) [47]:

[46] Compiled and calculated from the date of: "All-India Rural Debt and Investment Survey 1961-62. Tables Relating to Gross Farm and Non-Farm Receipts" (mimeo), pp. 3, 6, 23.

[47] J. Gutierrez, Rice Marketing in the Philippines, *vide*: "Economic Research Journal", vol. XII, 1965, No. 3, p. 204.

	% of gross harvest
Land rent in kind	17.45
Repayment of loans and interest	6.05
Sold or given in exchange	10.10
Total rice placed on market *:	33.60
Payment in kind for agricultural services	22.15
Consumed on farms	54.25
Total consumption in kind:	66.40

* Here it is assumed that all the rice received in the form of repayment of loans in kind enters the market.

Therefore, in this region rent in kind and repayment of loans in kind alone accounted for 70% of all the marketed produce (on the assumption that only one half of repayment of loans is transformed to marketed grain, this share drops to 67%, and the total rice produce placed on the market drops to 30.6% of the gross harvest). But if we remember that of the produce sold on the free market by the producers themselves, a considerable part is sold under compulsion (to repay usurious interest in cash, taxes, etc.), the percentage of marketed rice obtained by outright plundering and compulsion may rise to 80-85%.

Land rent in kind, however, remains, as was already pointed out, the most important component of the mass of rice which is ultimately sold on the market. In the Philippines the tenant population, though the operational land area they possess is less than half of the total, must carry the main burden of supplying the market with rice. On the contrary, farms of owner operators play a relatively insignificant role in the formation of the national marketed grain fund (see Table 9).

All the rice producers—landowners taken together, even though they accounted for 46% of the production, supplied only a quarter of the total marketed grain in the country, while farms which to one degree or another are linked with semi-feudal (almost exclusively share-cropping) leasing of land supplied the market with almost three-quarters of all the marketed rice. Moreover, the level of commercialized grain production on leased lands, as a result of semi-feudal oppression, was 2.5 times higher than that on the farms where the land was owned by the producer himself.

From the point of view of its economic nature, most of the marketed produce coming from petty farms—both "freely" sold and squeezed out by landlords and money-lenders—is the result of unpaid surplus labour—and what is especially important—part of the necessary labour of the producer. Moreover, the concrete form and stages by which this produce is sold on the market (for instance, whether it is sold on the market by the producer himself with its monetary equivalent then appropriated by various pre-capitalistic layers or by the state, or, on the contrary, it is first appropriated by the latter in kind, and then realized on the market as exchange value) are, in principle, of little consequence.

Table 9

Distribution of marketed rice among different groups of farms on the Philippines (1956-1962) *

Producers by tenure status	Share in total number of farms, %	Share in total operational land area, %	Share in total rice produce, %	Share in total marketed rice produce, %	Marketed rice produce as % of gross rice output
1	2	3	4	5	6
Full landowners	46.0	61.3	45.2	24.5	16.9
Landowner-tenants	24.9	20.8	12.6	15.5	38.4 (24.5)
Full tenants	29.1	17.0	41.4	58.9	44.4 (36.3)
Farm managers on lands of big proprietors	**	0.9	0.8	1.1	41.7
Total	100.0	100.0	100.0	100.0	31.2

* Calculated and compiled on the basis of: "Economic Research Journal", vol. XII, 1965, No. 1, pp. 26-27 (1962 survey; columns 2-3); *Ibid.* vol. III, 1957, No. 4, pp. 205, 211-212 (1956 survey; columns 4-6).

** Less than 0.1%.

Note: The figures in brackets in column 6 represent the share of land rent and payments in kind for debts to the landlord.

Thus, the marketed food balance in a number of developing countries seems to rest on two cardinally opposite socio-economic foundations: while the level of a nation's requirements in marketed food is determined by the extent of the social division of labour, by the growth and strengthening of economic structures based on commodity production, the real satisfaction of these requirements still depends, by and large, on how and to what extent the pre-capitalistic forms of land ownership and capital affect the petty producer.

Since it is not the actual output, but the real consumption which determines the share and the mass of grain that will come to the market from the petty farms, wherever there appear any socio-economic factors (or any combination thereof) which tend to ease the pressure on the producer from pre-capitalistic forms of oppression, the tax burden and so on, a universal tendency crops up—the petty producer supplies less grain to the market. And this already has a marked effect on the dynamics of general, national mass of marketed grain. For instance, when the Pakistan rural communities got rid of the oppression by "professional" Hindu moneylenders (after 1947 they fled to India), who controlled a significant part of the assets of the nation's rural debt, this should have inevitably led to a certain growth in the grain consumption *in natura* by the peasant households with a corresponding reduction in the marketed mass of grain. Certain countries (for example, the UAR) succeeded in reducing rents by implementing land reforms. This also resulted in a higher grain consumption in the rural areas, notably, at the expense of that share which had previously flown to the urban market.

It should be noted that in those cases when the agrarian reform entails a real re-distribution of big semi-feudal land property in favour of the direct producers, it considerably expands and consolidates the petty production structure, and in so doing, as an inevitable result, triggers a considerable growth in the food consumption in the peasant households. In other words, a growth in petty land property, when it becomes a stable tendency, at the initial stages, leads to a certain "naturalization" of agricultural production. This is only natural because a land reform which changes relations of ownership, first of all, changes the nature of distribution of the product; its influence on increasing the volume of output of this product, especially when it concerns petty farms, is felt later. And if this process of "naturalization" is accompanied by a further parcelization of operational land holdings and by a reduction in the average production units possessed by the great majority of producers (due to the natural growth of the rural population and the resulting further drubbing of petty farms) the subsistence nature of petty farming is especially vividly expressed.

The UAR is quite indicative in this respect. In the period between 1900 and 1952 the average size of a land holding in the country fell from 5.6 feddans [48] to 2.1 feddans.

[48] 1 feddan = 0.42 hectare.

And in 1965 this figure stood at 2.0 feddans. What is more, while in 1900 the petty owners (up to 5 feddans) held only 21.8% of the land, in 1952 the corresponding figure was already 35.4%. There was a corresponding drubbing in the average units of cultivated land. The land reforms of 1952 and 1961, whereby part of the land was taken away from the landlord class, increased the share of the petty land property to 57.1% of the total by 1965. [49] Even though the average size of the holding in this group of owners increased somewhat (1.5 feddans in 1900, 0.8 feddan in 1952, 1.2 feddans in 1965), it was absolutely insufficient for creating any normal marketed surplus of food. Petty owners, previously tenants, who received land under the reform, naturally increased their own food consumption and almost completely stopped deliveries of commercial grain to the national market. Moreover, they make up a large contingent of the population that obtains a significant part of its food fund from the market. In our opinion, it was the positive influence of the land reform, which resulted in a re-distribution of a large slice of the national income in favour of the working section of the rural population, that was one of the main reasons for quite a pronounced increase (22.4% in the period from 1951 to 1963) in the average per capita food consumption in that country.

The dynamics of grain prices has a telling influence on deliveries of marketed produce by petty producers. Due to the subsistence basis of the economy, the flow of grain produced by petty farmers to the market is, as a rule, inversely proportional to the fluctuation of grain prices on the market. For instance, if there is a tendency for the market price to rise, the petty producers usually cut down their deliveries of grain to the market: in this case the petty producer has a chance to repay his monetary obligations (taxes, interest on loans, cash rent, etc.) by selling less grain than previously. [50]

---

[49] D. Warriner, Land Reforms and Development in the Middle East, London, 1957, p. 30; "Statistical Handbook of the United Arab Republic 1952-1965", pp. 46, 50.

[50] Concerning the relationship between marketed grain deliveries and fluctuation of grain prices, the Reserve Bank of India notes: "When prices... are high, the natural reaction of the farmers would be to increase their own consumption of the foodgrains they produce and to build up inventories of their produce with a view to realizing higher prices. The marketed surplus then would not increase at the same rate at which agricultural income increases,

Whenever there is quite a low level of marketability in grain farming, even the slightest increase in the food consumption of producer families has a very noticeable effect on the general flow of marketed grain to the market. In one of its reports, FAO presents a vivid example: if normally 25% of the total grain output is realized on the market and 75% is consumed by the producer families, then an increase of only 1/10 in the consumption of the producers (meaning that the consumption fund would rise from 75% to 82.5% of the total produce) would cut grain deliveries to the market down to 17.5%, which means a reduction of 30%.[51]

There was in recent years in India a sharp drop in the amount of grain delivered to local markets, following a rapid increase in grain prices. One of the reasons for this (but not the only reason, as we shall see later) was the reaction of the petty producer to fluctuations on the market. However, it is important to stress that the process of reducing grain deliveries to the market, in this case, was in no way connected with the dynamics of grain production. In 1964/65 India took in the highest rice and wheat harvests in the entire period of the third Five-Year Plan, but the deliveries of these grains to the local markets fell considerably (see Table 10).

However, even though the switch-over of grain farming from one economic system (petty production burdened by pre-capitalistic exploitation) to another (large-scale, highly commercialized economy of capitalistic, or at times, public type) in the majority of developing countries is still at the initial stage, the growing sector of large-scale production will inevitably introduce certain correctives into the overall picture of grain arrivals on the market.

Take India, for instance. At present, the country has a group of agricultural regions which yield a large mass of marketed grain. According to surveys, in 1965 out of several hundreds of Indian districts 58 accounted for 48% of the marketed grain supply in the country.[52] In some of them, the capitalist

---

Conversely, when the agricultural prices... fall, the farmers would tend to reduce somewhat their own consumption of the farm produce and further reduce their inventories; both these factors would tend to increase their marketed surplus" ("Reserve Bank of India Bulletin", March 1960, p. 321).

[51] "National Food Reserve Policies in Underdeveloped Countries", 1958, p. 10.

[52] "The Economic Times", Bombay, 1966, January 1.

Table 10

Arrivals of food grains to local markets in India *
(1961/62=100)

Year	Rice **		Wheat	
	Arrivals	Wholesale price	Arrivals	Wholesale price
1962/63	92.2	106	108.0	99
1963/64	73.3	119	103.1	109
1964/65	63.7	128	76.7	143
1965/66	52.8	134	80.5	152

* Calculated and compiled on the basis of: "Review of Food Situation. April 1965", Delhi, 1965, p. 16; "Agricultural Situation in India", August 1966, pp. 391, 392; "Reserve Bank of India Bulletin", December 1966.

** Rice arrivals for 1965/66 are our own estimate; obtained by comparing the similar data for arrivals in 9 months of 1964/65 and 1965/66. The data for 1961/62 concern 94 markets, and since 1962/63—84 markets. However, these 84 markets account for almost the entire mass of rice arrivals, which previously was sold on 94 markets (for 1962/63 and 1963/64—97 and 96% respectively). That is why the figures, on the whole, reflect the dynamics of rice arrivals.

structure which ensures a more or less steady growth in the production of marketed grain has already got a firm foothold. For example, in the district of Firozepore (Punjab State), the average annual grain output increased from 239,000 tons in 1951-52—1953-54 to 362,000 tons in 1959-60 — 1961-62, i. e., over 8 years the rate of increase was 5.25% per annum. In the same period, the grain deliveries to the district market (for the purpose of subsequent dispatch of the grain outside the district) more than doubled (growth of 113%), and in the overall output of wheat they amounted to 16.6% and 23.3% respectively for the given periods. In the pure increase of production between the two periods, the share of marketed surplus amounted to 32.4%. [53] It is apparent that the role of capitalistic farming in the formation of the marketed mass of grain has grown to a considerable degree. However, along with this process, in India, as in many other developing countries, there was another process taking place—towards further reduction (due to parcelization) in the average output of the overwhelming mass of producers. And all this, as was pointed out previously, objectively intensified the subsistence nature of petty production and en-

[53] "Agricultural Situation in India", April 1963, p. 7.

hanced the tendency (especially under the influence of the above-mentioned processes—rising market prices, etc.) to reduce marketed grain deliveries by petty producers.

The State of Kerala is very indicative in this respect. Here petty and small parcel farming accounts for an extremely large share of agricultural production, in particular, in grain growing. According to the 16th Round of National Sample Survey (1959-1960), the share of farms having less than 2 hectares of land amounted to 92%, moreover, this group of producers held 41% of all the farm land in the state. However, they realized on the market, as is shown by an enquiry conducted by the State Bureau of Economics and Statistics, only 2 to 6% of their rice produce. Producers having from 2 to 4 hectares of land sold 17% of their rice harvest on the market. In other words, the production of grain in these groups of farms, especially in the first group, has a pronounced subsistence basis. On the other hand, farms of 4 hectares and more (3% of all the farms) sold a relatively large part of their rice output on the market—38%.[54]

In these conditions, the ultimate quantity of commercial grain reaching the market is a sum balance of the above-mentioned opposite influences. And it is quite understandable that the predominance of this or that tendency in the marketed grain supply for any sufficiently long historical period will depend on whatever factor has the stronger influence. For instance, here are the changes that have taken place in the co-relation of gross wheat output and marketed wheat produce (second most important grain crop) in the countries of the Indian subcontinent in the past 25-30 years. Official surveys and statistics indicate that the level of commercialization in the wheat farming as a branch of agriculture had a tendency to fall.

For example, in the mid-thirties in two of the main regions specializing in wheat farming—the United Provinces (what is today Uttar Pradesh) and the Punjab — 45-50% and 40% of the gross wheat harvest respectively went to the market. In the fifties, in Uttar Pradesh, only 35% of the wheat output was sold on the market (and as little as 28% in 1955/56), while for the Punjab the corresponding figure was 36%. In the early sixties (according to data for 1960/61) the Pun-

---

[54] "Survey on the Distribution of Agricultural Produce", Trivandrum, 1961, p. 15.

jab let only 31.6% of its wheat harvest reach the market. A similar tendency was observed in other regions where wheat fields occupied quite a prominent place in the overall cultivated land area. For instance, on the territory comprising the present-day Rajasthan, the level of marketability in wheat farming in the course of two decades dwindled from 75 to 44%, while in Bihar it slipped from 66 to 30%. For India as a whole, the marketed output in wheat farming dropped from 55% in the mid-thirties to 33-37% in the mid-fifties. [55]

The same process occurred in Pakistan. It should be noted that at present the grain farming supplies the market with only 20-25% of its output, while the main wheat growing regions—Punjab and Sind—in the mid-thirties delivered 40 and 66% respectively of their total wheat harvest to the market.

The actual volume of marketed wheat produced on the territory of the subcontinent followed approximately the following pattern: [56]

Country	Mid-30's			Early 60's *		
	Average annual harvest, mln. t.	Marketed surplus, %	Marketed surplus, mln. t	Average annual harvest, mln. t	Marketed surplus, %	Marketed surplus, mln. t
India	no data available			10.5	37 **	3.9
Pakistan	no data available			4.1	33 ***	1.3
Total	9.4	55	5.2	14.6	36	5.2

* Average annual harvests given for the period: 1962/63-1964/65 in India; 1961/62-1963/64 in Pakistan.
** Highest national estimate.
*** Assumed estimate; exceeds the level of commercialization of the entire grain output by 1/3.

It is evident that even though there was a significant increase in wheat production in both countries (55% in 25-30 years), the absolute volume of marketed grain apparently re-

[55] "Agricultural Marketing in India. Report on the Marketing of Wheat in India", Delhi, 1937, pp. 15-16; *Op. cit.* (Revised edition), Nagpur, 1963, p. 21; "Agricultural Situation in India", March 1962, p. 1212; "Indian Agriculture in Brief", 1965, p. 163.

[56] "Agricultural Marketing in India. Report on the Marketing of Wheat in India", 1937, p. 16; "Economic Survey 1965/66", Delhi, app.; "Agricultural Situation in India", April 1963, p. 3; "Economic Survey of Pakistan 1963/64", "Statistical Section", p. 18.

mains more or less stable. Wheat farming is becoming more and more subsistent in nature, or to be more precise, the tendency towards subsistence economy in wheat production is much more intense than the trend towards commercial economy. The growing capitalistic structure in both countries cannot, thus far, fill the vacuum with its increasing marketed mass of grain—a vacuum resulting from reduced grain deliveries to the market by petty producers.[57] (It is, however, noteworthy that one of the significant factors behind the transformation of wheat from a primarily "money" crop—55% of the gross harvest in the colonial period was directed to the market—into a "subsistence" crop, lies in that, following independence, certain sections of the petty commodity producers, due to improvements in their economic positions, switched from low quality to high quality grains in their consumption. Changes in the qualitative structure of consumption represented one of the reasons that caused a reduction in the output of marketed wheat).

Therefore, the dynamics of the marketed mass of food products is in no way related to the change in the actual national demand for these products; it does not fit in with this demand, and what is more, the gap between these two values generally tends to broaden. Herein lies one of the chief reasons for such a sharp deterioration in urban food supply in the countries in the past decade.

There is a cardinal change in the pattern of grain supply in the countries which are traditional exporters of grain. In the 50's South Korea almost completely dropped out of the ranks of rice exporters and started to import large quantities of wheat. South Vietnam—the rice basket of the entire Indochinese Peninsula—in 1958-1960 cut its export deliveries of rice to 7.6% of its total output (while in the late 30's Vietnam together with Cambodia had delivered 30.3% of their gross rice harvest to the foreign market). In 1961-1964 the export of rice from South Vietnam comprised only 4.2% of its total output, while for Cambodia the cor-

[57] It is interesting to note, for instance, that in India certain authoritative official agencies, while estimating the expected growth of marketed rice (most important grain crop) for the coming five years (1966-1971), proceed from a fixed level of commercialization, determining it as only 31.4% of the gross annual harvest ("Report of the Agricultural Prices Commission on Price Policy for Kharif Cereals for 1965/66", Delhi, 1965, p. 23).

responding figure was 18.5%. On the whole, the countries of the Indochinese Peninsula, beginning with the second half of the 30's and up to the mid-60's, had cut their export of rice to the world market by approximately 3 times. The intervention of American imperialism in South Vietnam is seriously crippling agricultural production: harvests of the main crop—rice—are dropping drastically (in the period from 1961/62 to 1963/64 South Vietnam's average annual output of cleaned rice was 3.33 million tons, and in 1964/65 it amounted to 3.42 million tons; in the following two years, the harvests dwindled correspondingly to 3.18 and 3.04 million tons, or approximately by 10%), while the country itself is becoming an importer of food grains (in 1965 the country's import of rice amounted to 7.6% of its gross harvest for 1965/66). [58] According to the Economic Research Service of the U.S. Department of Agriculture, the per capita output of food crops in South Vietnam for 1966 was only 92 points (1957-1959=100). [59].

A very typical tendency is observed in Burma and Thailand — two of the main rice exporters in Asia (in 1961-1963 they accounted for 72% of all rice exports from the countries of Southeast Asia and the Far East and about half of the world export of rice). [60] In the past quarter of a century the share of exported grain in the gross harvest fell by almost one half. Actually, Burma's volume of exports fell drastically (almost halved), while in Thailand, after a sharp reduction in the 50's, it exceeded the pre-war level somewhat. However, the average annual rice harvest in Thailand more than doubled compared to pre-war times (see Table 11).

Meanwhile, the bulk of marketed grain in these two countries makes up the export surplus. On the whole, all the countries of Southeast Asia and the Far East during a quarter of a century (from 1934-1938 to 1958-1962) reduced their average annual export of rice to the world market from 9 million tons to 3.6 million tons, i. e., a drop of 2.5 times. [61]

---

[58] According to the data in: "Economic Survey of Asia and the Far East 1963", ECAFE, 1964, p. 103; "Monthly Bulletin of Agricultural Economics and Statistics", December 1966, pp. 19-21.

[59] "The Far East and Oceania Agricultural Data Book. May 1967", Washington, p. 5.

[60] "Agriculture in the Southeast Asian Rice Bowl and Its Relation to U.S. Farm Export", Washington, 1965, p. 32.

[61] "The State of Food and Agriculture 1964", p. 215.

Table 11

Burma and Thailand: output and export of rice
(thousand tons of cleaned rice—annual average) *

Country	Production			Export					
				1937-1939		1958-1960		1962-1964	
	1937-1939	1958-1960	1962-1964	actual volume	as % of output	actual volume	as % of output	actual volume	as % of output
Burma	4961	4619	4902	3013	60.5	1617	35.0	1623	33.1
Thailand . . .	2956	4769	6077	1498	47.3	1142	23.9	1546	25.4

* "Economic Survey of Asia and the Far East 1963", p. 104; "Economic Survey of Asia and the Far East 1965", pp. 274, 302.

All this indicates that in grain-exporting countries themselves the national grain production is increasingly geared to meeting internal demand. And here, caused by a host of reasons (drubbing of land, etc.), grain farming shows steadily growing consumption tendencies which sharply retard the growth of marketed grain produce, sometimes even to the extent of reducing it (as is evident in Burma). At the same time, this mass of marketed grain available is distributed in such a way that the share consumed within the framework of the national economy, due to the relative and absolute increase in population demanding marketed grain, increases constantly. (And this results in a sharp reduction of the export surplus: in 1966/67 the rice exports from Burma amounted to only 600,000 tons!). [62]

Therefore, in the overwhelming majority of developing Asian countries it is possible to trace one and the same ten-

[62] In recent years, however, certain exceptions have been observed in this general tendency. This concerns such countries as Thailand and Cambodia which, due to a growing demand for rice on the world market, are gradually stepping up their rice exports. For instance, Thailand in 1964 and 1965 exported 1,875 thousand tons of rice annually, against 1,419 thousand tons in 1961-1963 (an increase of 32%), while Cambodia in the same years exported correspondingly 478 thousand and 245 thousand tons of rice (a 95% increase). There was a certain increase in the share of exports in the gross output of rice (in Thailand — approximately up to 29% in 1964-1965). But the future will show whether or not these tendencies are powerful enough to restore the export potential of rice farming in these countries.

dency—under the existing structure of agriculture there is a trend to stifle the growth of the mass of marketed grain in accordance with the growing demands for it. These demands in different countries arise due to different concrete circumstances: in some countries—caused by the need to, first of all, ensure the national food consumption fund (for instance, India, Pakistan, Indonesia and other importing countries), in others—actuated by the need to increase the foreign exchange components of the savings fund without which it is impossible to achieve expanded reproduction in the national economy (for instance, Burma, Thailand and other grain-exporting countries).

It is necessary, however, to stress that, of course, it is not the increase in consumption of grain *in natura* by certain groups of producers as such that stunts the growth of marketed grain supply, causing acute disproportions in the development of national economy, but rather the crisis of the entire socio-economic structure of agricultural economics and the preservation of class forces and production relations in rural areas preventing the growth of savings in agricultural production.

Due to the growing shortage of marketed grain, speculative capital is becoming increasingly active in commercial grain dealings in a number of countries. Taking control of the market, speculating merchants hoard marketed food products at their warehouses for long periods. In so doing, they disrupt the entire cycle of the food movement from the producer to the consumer and artificially intensify the shortage of marketed grain. Such actions by speculative capital place the process of labour power reproduction at a great disadvantage (holding back transformation of wages into consumer goods, the appropriation of a certain part of the wages by the speculators, etc.). And all this takes place at a time when the actual standards of consumption for the majority of the nation are considerably below the necessary minimum.

Huge stocks of marketed grain are concentrated in the hands of a special group of speculator merchants—the big land-owning class which survived the reforms. These rich landowners amass marketed grain not only by stocking up their granaries with grain they receive in the form of rent, or coming from their own farms, but also by buying up grain from the petty "independent" grain producers who are shackled by the money-lending capital of such landowners. As a result of such operations by merchants and rich landowners, the shortage of food grain for current consumption acquires such proportions in certain countries as to develop into a real national calamity. It is not only stagnation of grain production and insufficient deliveries of marketed grain from rural areas, but also the intensive speculation in food trading that explains why in many vast areas in India with scores of millions of population there developed near-famine conditions several times in the 60's. Speculators of all kinds controlling fabulous food stocks today have interwoven themselves organically into the social structure of Indian society, forming a close-knit parasitic hierarchy recruited from among the richest sections and groups of the exploiting classes. And it is, above all, monopolistic capital and its banks, supporting speculative tendencies in food trading, that rake in the greatest profits from such a situation. [63]

THE GROWTH IN FOOD IMPORTS

One of the most acute socio-economic problems in the developing nations, arising from the insufficient reproduction of food resources is the search for ways to regularly replenish, in the necessary amounts, the national food fund to comply with the demands for food which grow by the year. That model for socio-economic development which the ruling classes in many developing Asian countries have been proposing so far (intensifying economic growth without any effective changes in the old agrarian structure, clearing the way for all-embracing transformations in the entire system of agricultural production; relatively small capital investments in this sector of the national economy and in the branches of industrial production serving it) hampers the solution of the given problem on a national basis in the near future. So far, the tendency has developed in reverse: every year the food resources deficit is getting more and more alarming.

Under the circumstances, the only way to stop the ever-increasing gap in food supply remains to "count on the for-

[63] According to the Ministry of Food and Agriculture, Government of India, bank loans given to speculators only for hoarding wheat grew fourfold over the period from 1958/59 to 1963/64 ("Economic Survey of Asia and the Far East 1964", p. 122).

Table 12

Grain imports to the countries in the region of Southeast Asia and the Far East and in the region of the Near East

(million tons — average per annum) *

Region **	Type of grain	1934-1938	1948-1952	1953-1957	1958-1962	1963	1964	1965	1966***
I. Southeast Asia and the Far East . .	all types	7.4	8.9	10.6	14.5	19.9	23.0	25.0	27.7
	wheat	1.0	4.9	5.6	8.8	11.5	13.3	14.5	15.2
II. Near East	all types	0.5	1.8	2.1	4.5	5.2	4.5	4.9	5.3
	wheat	0.3	1.4	1.7	3.5	4.4	3.4	4.2	4.5

* "The State of Food and Agriculture 1964", pp. 215, 217; "The State of Food and Agriculture 1967", p. 183.

** Region I includes: Pakistan, India, Burma, Ceylon, Indonesia, Malaysia, Thailand, the Philippines, South Korea, as well as Taiwan. Region II includes: Iran, Iraq, Israel, Syria, Turkey, and certain other smaller countries of Asia Minor, as well as the UAR.

*** Preliminary estimate.

Note: A large part of the import fell on Japan which imported (with fluctuations for certain years) in the 50's and 60's approximately 3-4 million tons of grain.

eign market". As statistics reveal, in the post-war years the developing countries have had a tendency to steadily increase their imports of food (primarily grain) — see Table 12.

At present, a number of countries cover a significant part of their current demands for marketed grain by imports. (It is the imported grain that makes up for the shortage in the national marketed grain produce.) On the basis of the data on hand, it is possible to determine, for instance, that in India in 1961-1965 (with a mean output of marketed grain amounting to 30-32% of the gross harvest), the imports covered more than 18% of all the demands for marketed grain that were met. Moreover, the proportion of imports in India's gross marketed grain is steadily rising. Really, with the stagnancy in grain production in the first half of the 60's, the country imported increasing amounts of grain from year to year. While in 1961-1963 it imported an average of 3.86 million tons of grain a year, in 1964 the grain imports totalled 6.27 million tons; 7.6 million tons in 1965 and in 1966 reached the figure of 10.2 million tons. On the average for 1965 and 1966 the mass of imported grain made up approximately 30% of the nation's total marketed grain fund.

In Pakistan from 1960/61 to 1964/65 the imported mass of grain (with an average national level of marketability of grain farming amounting to 25% of the gross grain output) satisfied 26.5% of the nation's current demands for marketed grain, while in 1950/51-1954/55 the net import of grain (import minus export) accounted for not more than 4% of marketed grain consumed in the country. Just as India, so Pakistan in recent years has been compelled to sharply increase its imports of grain. In 1966/67 Pakistan imported about 2.8 million tons of grain against 1.6 million tons (annual average) for the first five years of the 60's. This means that the proportion of imports in the marketed mass of grain available within the country (if we take the grain harvest for 1965/66 as the basis) will rise to 40%.[64]

The Philippines, too, are importing more grain. In the first half of the 60's the proportion of imported rice—the main staple—accounted for approximately 24% of all the marketed rice consumed in the country. (Rice farms on the Philippine Islands, as we have seen, channel approximately 32% of

---

[64] According to: "Economic Survey of Pakistan 1965-66", Rawalpindi, 1966, "Statistical Section", pp. 10, 12, 39; "Economic Survey of Pakistan 1966-67", 1967, "Statistical Section".

their output to the market). Moreover, in 1965 imported rice covered 40% of the national demand in marketed rice (rice crop of 1964-1965 taken as the basis). [65]

In certain countries including Ceylon and the UAR the proportion of marketed grain coming from imports rises still higher. Unfortunately, there are no statistics available indicating how the UAR's national fund breaks down into marketed grain and grain produce consumed by the households of the producers and not reaching the market. It is, however, known that the productivity of labour in agriculture in the UAR is higher, for example, than in India or Pakistan. With all other factors being equal, this presupposes a higher level of commercialization in farming. However, as was previously pointed out, due to the strengthening petty commodity production structure, developing at the expense of large land holdings and the resulting semi-feudal forms of economy, there has been a noticeable increase in the consumption of grain *in natura* by the households of petty producers. This factor led to a reduction in the output of marketed food products. If we assume that in 1961-1964 the market received 30% of the grain produced in the country (the gross harvest in those years amounted to 4,725 thousand tons) and taking into account the net import of all food grains (in 1962-1965 the UAR imported 2,162 thousand tons of wheat, wheat flour—evaluated in terms of wheat—and maize and exported 345,000 tons of rice), the result will be that the proportion of net grain imports in the gross supply of marketed grain in the country amounts to 56%. If we take such a high ratio as 40:60 as the relationship between national marketed and non-marketed grain produce, in this case the share of imported grain will be equal to approximately 47% of the nation's gross supply of marketed grain. [66]

Therefore, one of the typical features of the food situation in many developing countries is that the exchange of the total

[65] According to: "Monthly Bulletin of Agricultural Economics and Statistics", December 1966, pp. 19, 21; "Economic Research Journal", Manila, vol. XII, 1965, No. 1, p. 14; *Ibid.*, vol. XII, No. 3, 1965, p. 204.

[66] According to statistics: "National Bank of Egypt Economic Bulletin", Cairo, 1966, No. 4; "Statistical Handbook of the United Arab Republic 1952-1965", p. 32; "Monthly Bulletin of Agricultural Economics and Statistics", February 1967, pp. 30, 34. In estimating the gross grain harvests, the paddy harvest was recalculated for rice at the equivalent of 0.66.

earnings of the working population for food products extends to an ever greater degree beyond the framework of the national market. This exchange is carried out more and more through the sphere of the world (and what is especially important— the capitalist) market. This particularly concerns the food supply in urban areas where the most productive (in countries where the economy is not as highly developed) part of the population — the industrial workers — are concentrated. For instance, in India, according to a number of indirect statistics, from 1/3 to 2/5 of all grain consumed in urban communities in the first half of the 60's was formed from imports. And in Pakistan, the proportion was apparently even higher. This increasing dependence of the people's food supply on the outer market for a number of developing countries means a specific stage of their involvement in the system of world capitalist economy; for the highly developed capitalist countries this opens up new opportunities and new prospects for economic expansion in these lands, which, in addition to being markets for fixed capital and industrial raw material, become a vast market for raw farm produce. But, undoubtedly, the most important aspect here is that imperialism is actively capitalizing on this type of dependence to suppress democratic tenden cies in the developing countries; in a number of countries imperialism is making use of this dependence to step up the consolidation of the class forces which uphold the most reactionary forms of national capitalism — capitalism of a conservative type. This is precisely the social mission that the U.S.A. has geared its aid mechanism to within the framework of PL 480.

The U.S.A. has been practising its policy of penetrating into the sphere of food supply in the developing countries especially intensively in the past 12 years. In the 60's a large part of the total grain imports of the developing countries came from the non-commercial delivery of grain products from the U.S.A. In the years from 1962 to 1964 wheat deliveries — the main imported grain—(see Table 12) under PL 480 amounted to 56% in the countries of Southeast Asia and the Far East, and 71% in the countries of the Near East of all the imports of this product; and the sum total imported by the two groups of countries within the framework of the food aid programme amounted to 26.3 million tons of wheat.[67] In general, the grain deliveries under PL 480 in those years

[67] "The State of Food and Agriculture 1965", pp. 45, 245.

accounted for about one half of all the grain imports of both regions. In certain countries the percentage of non-commercial grain imports from the U.S.A. is especially high. In India it amounted to more than 70% (1956-1965), in Pakistan—over 80%. Moreover, this import satisfied quite a significant part of the current demands of both countries for marketed grain—no less than 13-14% in India and more than 20% in Pakistan. Thus, the growing vacuum in the food markets of the Asian developing countries is filled over and above by U.S. farm produce.

At present the programmes of U.S. governmental agricultural exports are increasingly oriented on the developing nations of the area with the biggest food shortage, namely, Southeast Asia and the Far East. At the end of the 1950's the area received 38-39% of all the U.S. exports under PL 480, while in 1964 and 1965 this share rose to 50 and 57% respectively. In 1965 and 1966 the countries of Southeast Asia and the Far East absorbed 30-31% of all American agricultural commodity (mainly food) exports, as compared with 21% in the mid-1950's [68]. Moreover, there was a considerable increase in the U.S. agricultural exports. According to available estimates, India alone consumed in 1965 and 1966 one-fifth of the gross wheat output in the United States and her share in the world exports of American wheat amounted to 29% in 1964-1966.[69] Hence, the food markets of the developing Asian countries are becoming increasingly important for the United States too. A wide system of U.S. economic and political influence on the developing nations is behind these figures.

As for the countries whose ruling classes are steering their socio-economic development along the capitalist path, the inflow of American grain, undoubtedly, eases to a certain extent the tense food situation thereby acting as a factor containing mounting popular indignation. And this, in turn, enables influential forces in the ruling circles and bureaucratic apparatus to hold back the transformation process in the agricultural system. This, of course, is in full accord with the interests of the big landowners who gain time for evolutionary changes of their vast land holdings to a system based on capitalistic principles. Thus, the U.S. food aid, helping to delay the land reforms, thereby reinforcing the position of

---

[68] "The Far East and Oceania Agricultural Data Book. May 1967", p. 53.

[69] *Ibid.*, p. 60.

those class groups which stick up for the conservative form of capitalism in agriculture (e. g.. capitalism of the "landlord" type), acts as a force slowing down the pace of development of productive forces in this branch of economy.

The United States readily resorts to such means of exerting pressure on the developing nations—recipients of U.S. aid, as "freezing" food deliveries under PL 480. It is this method that the U.S.A. employed, for example, with regard to India at the end of 1966. The reason, among others, for that action was to force the Indian Government to some important concessions in favour of the United States which is seeking to wrest one of the key industries, heavy chemicals, particularly the production of fertilizers, from the public sector and to establish the domination of the American oil monopolies in this branch.[70].

As shown by experience, the policy of orientation on increasing American food deliveries does not, by far, always prevent speculation by the rich bourgeoisie engaged in commerce and banking (including the monopolistic type) that grasp control of a significant part of the national resources of marketed food. However, this policy stems, to a certain degree, from the fact that the state is helpless to curb the speculative tendencies of this reactionary section of the bourgeoisie.

A noteworthy statement was made by a distinguished Indian economist, S. C. Gupta: "in view of these pernicious effects — actual and potential — of PL 480 imports and the dangerous possibilities to which they give rise, it is worth serious consideration whether a more active, vigorous and purposeful policy towards agricultural development would be preferable to the current strategy of making the country dependent on PL 480 imports."[71] It becomes all the more imperative to implement such a policy if we take into consideration that from the beginning of the 70's the U.S.A. is changing the system of payments for its agricultural deliveries to the developing countries to a dollar basis. This change will fully deprive the developing countries of the temporary benefits which prompted them to accept U.S. food aid (payment for PL 480 commodities in national currency). As a result,

---

[70] See: "Washington Post", 1965, December 8; "Amrita Bazar Patrica", Calcutta, 1966, November 22; "The Economic Times", 1966, December 21.

[71] S. C. Gupta, *Freedom from Foreign Food*,— "Pernicious Effects of PL 480", Delhi, 1965, pp. 52-53.

the foreign exchange components of the accumulation fund in the developing nations food importers, can be substantially drained and their low rates of economic growth would face the danger of being further slowed down.

THE MARKET AND REPRODUCTION IN GRAIN FARMING. THE INFLUENCE OF INTERNATIONAL FACTORS

The national economies in the developing countries are influenced in many ways by the dynamics of the world capitalist economy. The developing countries, for instance, make use, to a certain degree, although on very difficult terms, of the fruits of the rapid technological progress taking place in the developed capitalist countries. However, the world capitalist economy (through the system of state-monopoly capitalism and other "levers") with its inherent tendencies—to consolidate and increase the influence of its economic laws in the countries of the "third world"—acts in a number of cases as a brake, holding back the growth of the productive forces in the young national states. The depressive influence of these laws also affects the process of reproduction in the agriculture of the developing countries.

The preceding section dealt briefly with the tendency of increasing the dependence of a large group of emerging nations on the world capitalist food market in ensuring their national food supply. This dependence is refracted in the structure of the market ties of the developing nations, pointing, in this case, to the growing integration of national food markets with the world capitalist food market.

What is the economic mechanism whereby the world capitalist economy manifests itself as a "depressor" on the argicultural development of these countries? Here, as in the previous sections, we shall deal with questions confined mainly to grain production.

Capitalism in a group of highly developed countries during the past 15-20 years, as is known, was able to largely modernize agriculture on a machine-industrial basis. One of the specific features of this process was that, thanks to the revolution in science and technology that transformed argiculture in the post-war years, the growth of labour productivity in this branch of economy in a number of leading capitalist countries superseded that in industry.

For example, in France between 1950 and 1962 the productivity of labour in agriculture doubled; moreover, even today its growth rates still exceed the corresponding indices for other spheres of economy, amounting to 5.8% per annum against 4.6% for the latter.[72] A similar tendency was even more pronounced in the U.S.A. where from 1945 to 1958 the output of farm produce per man-hour grew by 124% (average annual rate of increase—6.4%)[73], and the total labour expended, for instance, on production of all crops dropped in the 1950-1964 period from 6.9 thousand million to 4.1 thousand million man-hours, or by more than two-fifths.[74] Moreover, there was a considerable increase in the total output of farm produce in this period. In certain branches of agriculture there were especially noticeable increases in the productivity of labour. For instance, in the period from 1960 to 1963 only 4.4 hours were expended on producing a ton of wheat in the U.S.A. while 15 years before it took 12.5 hours (according to the data for 1945-1949); or in the second half of the 40's the cultivation of one ha. of wheat required 14,3 hours, whereas at the beginning of the 60's only 6.3 hours were needed.[75] As a result, the capitalist sector in the agriculture of the developed countries (excluding, however, such areas as Japan, Southern region of Italy and some others), was rapidly approaching the level of industrial production by the labour productivity index. On the contrary, in the agriculture of the developing countries it was possible to observe a relatively stable average productivity of labour which, of course, was at an extremely low absolute level.

The tremendous gap between the levels of labour productivity in the agriculture of the developed capitalist countries, on the one hand, and the developing countries, on the other, reflected the increasingly uneven development of agricultural production in different sectors of the world economy. Compared to the countries of the West, the quantity of labour-time expended on a unit of produce in the developing countries ran

---

[72] L. Perceval, La question agraire en France,— «Economie et politique», Paris, № 139, Fevrier 1966, p. 58.

[73] «Экономическое положение капиталистических стран. Конъюнктурный обзор за 1963 г. и начало 1964 г.», М., 1964, стр. 57 ("The Economic Situation in the Capitalist Countries. A Statistical Review for 1963 and the beginning of 1964", Moscow, 1964).

[74] "Statistical Abstract of the United States 1965", Washington, p. 645.

[75] *Ibid.*, p. 646.

into astronomical figures. In the chief rice-growing countries of Asia 18-32 times more living labour was expended to produce a ton of rice than, for instance, in the U.S.A. The living labour required for producing a ton of wheat in India today exceeds by more than 100 times the same index in the United States of America (see Table 13).

Table 13

Living labour expenditure per unit of produce and output per one man-hour *

Crop	Country	Date of Survey	Labour expenditure per ton of produce (man-hours)	Output (kg/hr)
Wheat	India (Punjab)	1955-1957	492	2.0
	U.S.A.	1960-1963	4.4	227.3
Rice (cleaned)	India (W. Bengal)	1955-1957	947	1.1
	Thailand	1953	803	1.2
	Burma	1956	695	1.5
	Philippines	1958	535	1.9
	U.S.A.	1960-1963	30	33.3

* "Studies in Economics of Farm Management in the Punjab. 1954/55-1956/57", Delhi, 1963, pp. 79, 83; "Studies in Economics of Farm Management in West Bengal. 1954/55-1956/57", Delhi, 1963, pp. 94, 98; "Economic Research Journal", Manila, vol. VI, 1959, No. 2, p. 76. "Burma Agricultural Planning Commission", Rangoon, 1956, vol. VII; "Statistical Abstract of the United States 1965", p. 646; Esihiko Hasegawa, Tai-no Beimoku Jijo, Tokyo, 1962, pp. 223-224, 228-229; «Экономическое положение капиталистических и развивающихся стран. Обзор за 1964 г. и начало 1965 г.», М., 1965, стр. 60. "Economic Situation in the Capitalist and Developing Countries. A Review for 1964 and the Beginning of 1965", Moscow, 1965).

The growing gap between the levels of agricultural development in the industrially-developed countries and the emerging nations is vividly illustrated by comparing similar indices for Turkey and France — two countries with approximately the same area of arable land. Though Turkey possesses the largest tractor pool among the developing countries of Asia and North Africa, it lags behind, first of all, in the sphere of mechanization of agriculture (see Table 14).

In Turkey the main increase in the pool of farm machines took place in the period from 1949 to 1954, after which there was a sharp drop in the level of mechanized agriculture. Some

Table 14

Farm machines in Turkey and France *

Country	Tractors ** (thousands)			Grain harvesters (thousands)			No. of tractors per 1000 ha of arable land (1963)	Tractor H. P. per 1000 ha of arable land (1960)	No. of grain harvesters per 1000 ha of grain crops (1966)
	1949	1954	1964-1966	1949	1953	1963-1966			
Turkey	2.9	33.7	44.5 61.1	0.3	4.5	5.9 7.7	1.8	50	0.6
France	76.9***	211.8	900.0 1070.0	5.0	11.2	85.0 110.0	49.2	860	12.1

* «Мировая экономика и международные отношения», М., 1965, № 7, стр. 155—158 ("World Economics and International Relations", Moscow, 1965, № 7), «Экономическое положение капиталистических и развивающихся стран. Обзор за 1966 г. и начало 1967 г.», М., 1967, стр. 58 ("Economic Situation in the Capitalist and Developing Countries. A Review for 1966 and the Beginning of 1967", Moscow, 1967); "Monthly Bulletin of Agricultural Economics and Statistics", FAO, Rome, May 1966, pp. 1-3; "The Summary of Agricultural Statistics 1966", Ankara, 1966, pp. 4, 15, 16.

** All tractors over 8 H. P., for Turkey in 1966 over 10 H. P.

*** 1948.

new increase in the number of farm machines is observed since the mid-60's. And on the contrary, in France there was a steady growth in the utilization of machines (moreover, it is important to note that French farmers extensively use garden tractors which numbered 32,700 in 1949 and 140,600 in 1962). The relative gap in the level of farm mechanization between the two countries widened to tremendous proportions.[76].

This fact is confirmed also by comparing the changes in the structure of horse-power employed in agriculture (draft animals, tractors). In the period between 1951 and 1960 the share of animal horse-power (at the equivalent 1 horse = = 7 H. P.) in France dropped from 81% to 40%, or by more than one half (the respective figures for the U.S.A. were 20% and 9%, for Canada—40% and 18%, for the Federal Republic of Germany—74% and 24%), in Turkey it dropped from 91% to 87%, or less than by 5%.[77] It is not surprising, therefore,

---

[76] It should be noted that most of the tractors in France are 35-50-H. P. machines. In Turkey only about half of the tractors are more powerful than 35 H. P.

[77] "Monthly Bulletin of Agricultural Economics and Statistics", May 1966, p. 4.

that while in Turkey tractors worked only 15.9% of the cultivated land in 1963 and 20.3% in 1966 (as against 8.6% in 1950 and 14,6% in 1956) [78], in France most of the arable land was worked by power-driven machines. According to the data on hand, only due to mechanization the total expenditure of living labour in agriculture in France since the end of the war was reduced by one third; moreover, in crop farming it dropped by 50%.[79] The progress in capitalistic agriculture, resulting in a considerable increase in the productivity of labour, also had another effect. It squeezed part of the economically active population out of the sphere of agriculture. For instance, on the big farms of the Paris region 11.5 workers were needed per 100 ha. of land in 1920-1930, 10.3 in 1949-1950, and only 6.9 in 1960-1961.[80]

The contrast between the two countries is just as striking in application of other technical means. In 1962-1963 France, for instance, introduced 122.4 kg of fertilizers (in terms of 100% of nutritional content) per 1 ha. of arable land, while Turkey used only 2.7 kg.[81]

Along with the growing difference in technical equipment and techniques, there is an ever-increasing gap in the levels of agricultural yields in Turkey and France. In particular, this stands out quite vividly from the following statistics on wheat yields, wheat being the chief grain crop in both countries (100 kg/ha):

Country	1934-1938	1951-1955	1962/63-1965/66
Turkey . .	9.8	10.7	11.5
France . . .	15.6	20.8	30.5

[78] "The Summary of Agricultural Statistics 1966", p. 3. Turkish statistics estimates the whole area worked by tractors, proceeding from the assumption that 1 tractor works 75 hectares a year. This figure, in all likelihood, is much too high, if we take into consideration the standards of the developed capitalist countries with their much more powerful tractor pool. In the USA, for example, where 80-100 H. P. units are predominant in 1966 a single tractor worked 27 hectares; in France — 16 hectares (see: «Экономическое положение капиталистических и развивающихся стран. Обзор за 1966 г. и начало 1967 г.», стр. 58). Therefore, in Turkey actually much less of the cultivated land was worked mechanically.

[79] L. Perceval, *Op. cit.*, p. 58.

[80] *Ibid.*

[81] «Мировая экономика и международные отношения», 1965, № 3, стр. 156.

While there was almost no change in the yields of crop farming in Turkey, in France it doubled within a quarter of a century. Moreover, in the last decade alone—a period of intense modernization in agriculture—it increased by almost 50%. The ratio of yields previously expressed by 1 : 1.5 now stands at 1 : 3.

Therefore, while in one group of countries (developed capitalist states) the main trend in agriculture was towards reducing the manpower employed along with an ever-increasing productivity of land and labour, in another group of countries (developing nations) the average national productivity of labour in agriculture was in a state of stagnancy rather than increasing. This means that in the past 10-15 years in a large part of the world, which (according to UN terminology) represented the system of market economies (the developed capitalist and developing countries), there was a radical change in the ratio of national indices for the labour-time required to produce one and the same unit of agricultural produce. There was also a significant growth in the quantitative difference between the national costs of production and, correspondingly, the national values of these products.

In other words, within the framework of the world "market economies" the agriculture in the industrially-developed countries counterposes, to an ever-increasing degree, the agriculture in the developing countries, just as large-scale highly-productive industry stands opposite petty, primitively-equipped farming.

At the same time, the scientific-technological revolution in the agriculture of the leading capitalist states was accompanied by an increase in the total output of food products, by the formation in a number of countries of colossal "surplus" food stocks,— which could not be sold either on the domestic or foreign market. There was an especially great over-production of grain in the capitalist countries. For instance, in 1952 the U.S.A.'s closing stocks of wheat amounted to only 7 million tons, by 1962 this figure already stood at 36 million tons,[82] i. e. it exceeded the annual output and was double the annual total consumption of this crop within the country. The unsold, i. e., "surplus" stocks of wheat in five countries—the chief grain exporters in the capitalist world (the U.S.A., Canada, Australia, Argentina, France)—amounted to 49

[82] "The State of Food and Agriculture 1965", p. 227.

million tons, 50 million tons, 42 million tons and 43 million tons respectively, for the middle of 1962, 1963, 1964 and 1965. Together with other kinds of grain (rye, maize, barley, millet, sorghum and oats) the grain stocks in these years amounted to as much as 100-120 million tons, moreover, most of these stocks (over 75%) were concentrated in the country with the highest developed agriculture—the U.S.A.[83] A very interesting picture unfolds before us if we compare the scale of overproduction of grain in the capitalist countries with the narrow scope of grain production in the developing nations. For instance, the "surplus" stocks of wheat alone in the five above-mentioned countries exceeded by 30-40% the total harvests of this crop in the developing countries of Asia and the UAR for 1962-1964, while the accumulated stocks of all the grains make up 80-90% of the total annual harvests of the main grain crops in the developing nations of the Southeast Asia and the Far East region which in 1965 was inhabited by 41% of the population of the capitalist and developing countries combined.[84]

On the whole, the "surplus" stocks of wheat alone accumulated in the 5 capitalist countries, were approximately equal to the average annual wheat export from these countries (exports of wheat amounted to 39 million tons in 1961/62, 38 million tons in 1962/63, 53 million tons in 1963/64 and 48 million tons in 1964/65). In other words, of the total stocks of wheat that were not consumed in the course of a year within these countries, only every second ton of grain could be sold on the outside market. The flooding of the world market with grain was quite formidable and steady—as a result of overproduction of grain in the developed capitalist countries.

Here we can see an economic law come into play, the law according to which once the market is overflooded, i. e., when the commodities supplied exceed the effective demand, the market prices at which the commodities are sold start to be regulated by the value of those commodities which are produced in the best conditions, i. e., at the highest productivity of labour and the lowest cost of production.

---

[83] "Monthly Bulletin of Agricultural Economics and Statistics", May 1966, pp. 10, 12.

[84] Compare: "The State of Food and Agriculture 1965", pp. 220-221; "Monthly Bulletin of Agricultural Economics and Statistics", September 1967, p. 11.

As for the world grain market,[85] this means that the international value of grain produce is determined today almost exclusively by the labour-time expended on grain production in the developed capitalist countries, and in the first place, in such countries as the U.S.A. and Canada — the leading grain exporters; and this labour-time, as we have seen from the example of living labour expenditure, is being steadily reduced in the course of the scientific-technological revolution in agriculture.

The situation in the world grain market has started to change considerably in recent years. In view of the great demand for grain in the developing nations, the unsold stocks of grain in the capitalist countries have begun to deplete gradually. According to a FAO estimate, the unsold stocks of wheat in the 5 countries previously mentioned shriveled to only 29 million tons in the middle of 1966, with the U.S. stocks to less than 15 million tons (as against 38 million tons in 1961 and 22 million tons in 1965). And the increasing demand for grain in the developing countries, if their output does not grow, can, in the near future, in general, devour these "surplus" stocks. However, since the developed countries remain the main exporters of grain on the world market (apparently, the U.S.A. will, for a relatively long period of time, persist in its role of the leading supplier of grain to the developing nations), the strings of the price regulation mechanism on the world capitalist grain market (with all possible price fluctuations) will still be pulled by the developed capitalist countries.

The cardinal question arising from this situation is the following: in what way does the tendency for the international value of grain to dwindle affect the process of price formation on the grain markets of the developing nations? Numerous facts indicate that this influence is of a depressive nature. Moreover, such an influence is all the greater, the more closely the grain farming and grain market of a given developing nation is tied to the world capitalist economy and its market.

---

[85] The play of this law in international trade is shown by N. Shmelev in his work: Н. Шмелев, Мировая торговля и освободившиеся страны,— «Вопросы экономики», М., 1964, № 8, стр. 80-85 (World Trade and the Liberated Countries, *vide*: "Questions in Economics", Moscow, 1964, No. 8).

As a result of the laws of world capitalist economy coming into play, the tendency in countries of this type is to push down the market prices on the national grain produce. Depending on the national situation (for instance, with inflation on the upswing and the resulting growth of nominal prices), this tendency may manifest itself in the grain price increase lagging behind the pace of price growth for other groups of goods.

In this respect, Turkey presents a really striking example. There was a significant drop in the prices on wheat — the chief grain crop in the country. In 1952-1954 the prices on soft wheat at the Istanbul Exchange equalled 11.8-12.5 dollars per 100 kg, whereas in 1961-1965 they fluctuated within the limits of 8.6-9.5 dollars per 100 kg. Within a decade, the real price of wheat, in terms of the "gold lira", fell by approximately a quarter (while the nominal prices, in terms of the lira banknote, increased almost 2.5 times).[86] The waning tendency of Turkish wheat prices is reflected in Chart 1.[87] It should be noted here that the national Turkish price in the 60's came close to the level of grain prices on the world capitalist market. For example, CIF prices on imported wheat delivered to Austria in 1961/62-1964/65 ranged from 8.5 to 8.7 dollars per 100 kg.[88] Thus, the high prices in Turkey existed precisely at the time when the European countries were completing the post-war re-establishment of their agriculture and were beginning to fully modernize it technically, with Turkey

---

86 "Production Yearbook 1964", p. 331; "Monthly Bulletin of Agricultural Economics and Statistics", March 1966, p. 48.

87 Charts 1 and 2 are compiled according to: "Production Yearbook 1964", pp. 330-331, 338-339; "Monthly Bulletin of Agricultural Economics and Statistics", March 1966, p. 48; *Ibid.*, September 1966, p. 52; "International Financial Statistics" (Supplement to 1963-1964 issues), Washington, 1963, p. X; *Ibid.*, November 1965, p. 27; *Ibid.*, May 1966, p. 28.

Since the charts are based on price indices in terms of dollars, it is necessary to point out that the extremely wide amplitude of Turkish prices in the period of 1956-1959 can be explained by the strong pressure of inflation tendencies (from 1954 to 1957 the money in circulation swelled from 1.9 thousand million to 4.1 thousand million Turkish lira—see "TCMB Aylik Bülten", Ankara, 1966, No. I, p. 45) and by compulsory maintenance of a high rate of lira-dollar exchange (curve going up—to 1957) and the subsequent devaluation of the lira (curve dropping sharply — to 1959). Comparable periods are the beginning of the 50's and the first half of the 60's.

88 "Prices of Agricultural Products and Fertilizers in Europe 1964-65", FAO, Geneva, 1966, p. 17.

Chart 1

The Dynamics of Market Prices on Wheat 1950-1965
(1952-1954=100)

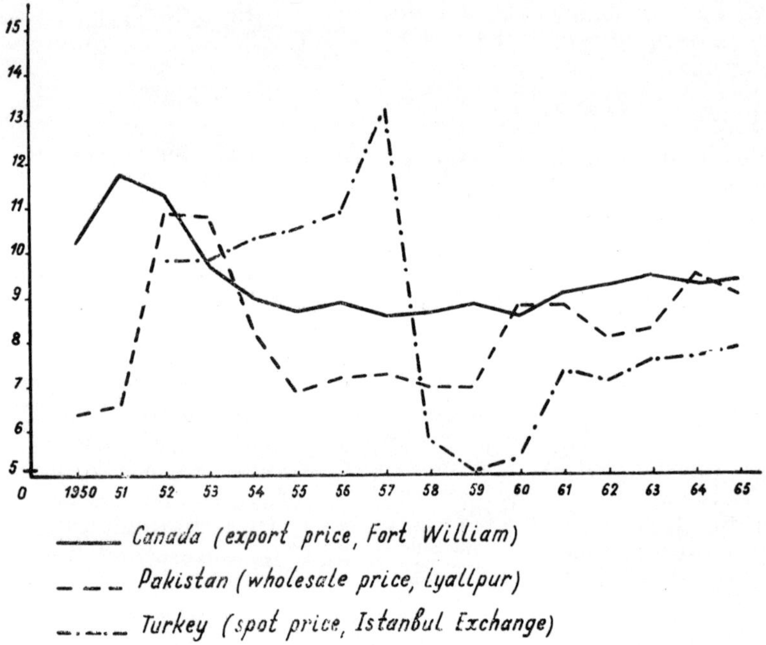

(up till 1954) delivering large consignments of grain to the continent. On the contrary, prices on national wheat dropped when there was a sharp over-production of grain in the developed capitalist countries, and Turkey began to feel the grain deficit and started to import large quantities of overseas grain.[89]

At the same time, naturally, there was a drop in the prices which the Turkish producers incurred when selling their pro-

[89] For details on this question, see: Г. И. Старченков, Воздействие военных блоков НАТО и СЕНТО на сельское хозяйство Турции,— «Проблемы экономики и истории стран Ближнего и Среднего Востока», М., 1966, стр. 122—125. (G. I. Starchenkov, The Influence of the NATO and CENTO Military Blocs on the Agriculture of Turkey, *vide*: "Problems of Economics and History of the Countries in the Near and Middle East", Moscow, 1966)..

ducts. They "pulled up" to the level of prices at which the West European farmers were selling their produce. It is interesting to note that the West German farmers, for example, thanks to the traditional policy of agrarian protectionism practised by the German capitalist state (which, as is known, caused serious friction in the Common Market between the Federal Republic of Germany and France which is a major exporter of agricultural produce), received even more than the Turkish producers for their goods (see Table 15). And all this took place with a great difference in the levels of the average labour productivity existing between Turkey's agriculture, on the one hand, and the West European states, on the other.

Table 15

*Prices on wheat realized by producers in France, FRG and Turkey **
(U.S. dollars per 100 kg, for all sales and all types of wheat)

Country	1961/62	1962/63	1963/64	1964/65
France	8.08	8.52	8.49	8.29
FRG	10.43	10.54	10.54	10.70
Turkey	8.02	9.16	9.12	9.04

* "Prices of Agricultural Products and Fertilizers in Europe 1964/65", Annex, p. 4.

The peculiarities reflecting the dynamics of prices on the main groups of commodities confirm the above-mentioned tendencies. As was already pointed out, when there are strong inflation tendencies (as is the case in Turkey and in a number of other developing nations), the nominal commodity prices rise continuously. However, in Turkey, in respect to different groups of commodities, the prices rise unevenly (see Table 16).

The growing prices on wheat lag behind those for the whole group of grains; the price on wheat also lags behind the growth of prices on industrial raw products and semi-finished goods. It is in this persistent gap in the relative price levels that the depressive influence of the world grain market on the national wheat market is manifested. In the economic sense, this gap reflects one of the forms of non-equivalent exchange between the agricultural sector of the economy and the industrial

Table 16

Indices of wholesale prices in Turkey *

(1953=100)

Year	General Index	Industrial raw products and semi-finished goods	Grain	
			All types	Wheat **
1955	119	128	105	105
1962	260	261	270	248
1963	271	271	268	227
1964	269	276	265	239
1965	293	296	297	257
1966	306	302	304	270

* "TCMB Aylik Bülten", 1967, No. 4-5, p. 58; "Production Yearbook 1964", p. 331; "Monthly Bulletin of Agricultural Economics and Statistics", September 1967, p. 45.

** Index calculated according to the price indices for soft wheat at the Istanbul Exchange.

sector, irrespective of the genetic nature of the latter (national production of industrial branches of the highly developed capitalist countries, whose manufactured goods enter the domestic market of the emerging nation through trade channels). This subject will be taken up in somewhat greater detail later on.

The fact that the cost of production of grain (plus the average profit) in the highly-developed capitalist countries, through the mechanism of the world capitalist market, becomes a regulator of the national market grain prices in a number of developing countries, retards the development of the productive forces in agriculture in these countries. And this is only natural, what with the average market prices received by the producers for their products falling faster than the growth of national productivity of labour and the national cost of production dropping. The result of such tendencies is a sharp retardment in the whole process of capital accumulation in agriculture.

In certain developing nations, the non-national regulator of national market prices acts with such a force that it not only obstructs the development of peasant forming, which has the lowest productivity of labour, but also holds back the growth of the productive forces on big farms, i. e., in the capitalist structure which is characterized by a higher (at any rate—higher than the national average) productivity of labour. As a result, the process of mechanizing Turkish agriculture which developed very turbulently at the end of the 40's and early in the 50's, came to a standstill exactly at the time when Turkey was ousted from the European grain markets and national grain prices sharply dropped. It is not surprising, therefore, that the rate of growth of agriculture, especially that of grain production, has diminished several times. Actually, in the 7 years (1946/48-1953/55) coinciding with the period of intensive mechanization of agriculture, the grain harvests in Turkey increased by an annual 7.3%. But in the following decade (1953/55-1963/65), they increased by less than 2.5% per annum.[90] The volume of grain output at the end of the 50's and first half of the 60's actually remained at the 1953 boom level.[91]

Naturally, the sharp drop in the rates of growth of grain production to a considerable extent prevents the country from resolving its food problem on a national basis. (Here we shall not deal with the fact so widely covered in Soviet literature that the ever-increasing tendency in the evolution of Turkey's agriculture towards capitalism of landlord type, which envisages lengthy preservation of the old forms of land ownership and capital, in itself hampers the growth of productive savings accumulation in the rural areas and the increase of agricultural production).[92]

Of course, the laws governing the world capitalist market have a depressive effect in the developing countries on the prices, above all, of those food-stuffs which are produced in surplus quantities, exceeding the effective demand, in the developed capitalist countries. Of the grains, in the first place, this concerns wheat. Meanwhile, the over-production of wheat in the developed countries, with the resulting tendency of preserving world prices on wheat at a level lower than the national prices, gives the developing countries — food importers — a chance of replacing more "costly" grain (in

---

[90] Calculated on the basis of data in "The Summary of Agricultural Statistics 1966", p. 4.

[91] "TCMB Aylik Bülten", 1966, No. 4-6; *Ibid.*, 1967, No. 4-5.

[92] See: Ю. Н. Розалиев, «Классы и классовая борьба в Турции», М., 1966, стр. 18-23 (Yu. N. Rozaliev, Classes and Class Struggle in Turkey, Moscow, 1966); П. П. Моисеев, Аграрные отношения в современной Турции, М., 1960, стр. 113-132, 197 (P. P. Moiseyev, Agrarian Relations in Turkey Today, Moscow, 1960).

Chart 2

The Dynamics of Market Prices on Rice, 1950-1965
(1953-1957=100)

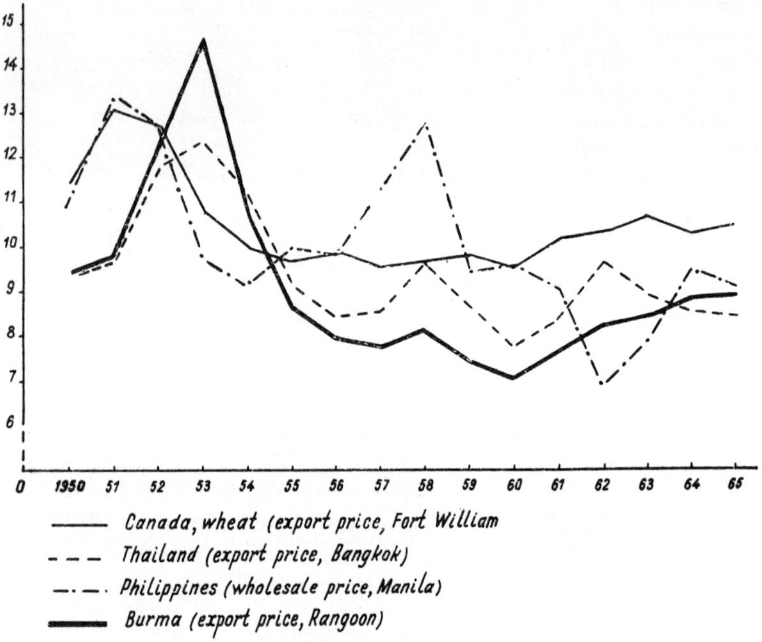

this case rice, which is the staple food for the people in Asia and the Far East) with a "cheaper" type of grain, i. e., wheat. This opportunity is backed by certain changes in the structure of the demand for food (for example, the growing urban population in a number of countries gives rise to an increasing demand for a "worse" type of grain for these countries — wheat, as compared to the demand for rice).

The resulting competition between the commodities of kindred branches in the end develops into a situation where the production cost and, ultimately, the value of the "substitute" commodity, i. e., wheat, in the developed capitalist countries acquires a regulating influence on the world market prices of the more "costly" commodity—rice, the bulk of which, as is known, is produced in the worst conditions—

with a low productivity of labour. To a considerable degree, this very factor is behind the general waning-price tendency for the rice supplied by the developing countries—the exporters of rice, the tendency that developed up till the mid-60's. For one and the same unit of rice exports, in the second half of the 50's and the first half of the 60's, these countries received on a yearly average less than in the first half of the 50's.

The dependence of world rice prices on world wheat prices, with a certain coincidence in the general trends of their movement is shown in Chart 2. Note the relative stability of the wheat prices as compared to rice prices, for which great annual fluctuations are typical, as a result of changes in the ratio of the current regional demand and supply. But the broad amplitude in the fluctuation of prices on rice only reaffirms the fact that the world price on wheat, as a kind of "invariable constant", serves as a centre of gravity for the world prices on rice. However, the relatively regular demand, stemming from century-old traditions of national consumption ("costly" rice is preferable to other kinds of grain), kept the world prices on rice from falling still further.

In recent years, however, in view of the formation of a colossal grain deficit in a large group of developing nations, the demand for all kinds of grain, including rice, grew considerably on the world market. "In the long run", the International Rice Study Group noted, "current trends suggest a growing gap between the supply of and demand for rice in developing importing countries unless steps are taken to accelerate the expansion of production".[93] These tendencies boost world prices on rice (they went up already in 1966).[94] The problem is whether or not the developing nations—rice exporters can make use of this opportunity, and if they can, for how long, in order to improve the conditions for reproduction in their agriculture.

The changes in the economics of rice farming in the South of the U.S.A. puncture any optimism in this respect. First of all, the U.S.A. is expanding its rice production at an ever-increasing pace. Moreover, rice farming is developing exclu-

---

[93] "Monthly Bulletin of Agricultural Economics and Statistics", December 1966, p. 18.

[94] For example, in 1966 the export price on Thailand's white rice was 16.6 U.S. dollars per 100 kg as compared to 13.7 dollars in 1964 and 1965 (FOB price in Bangkok). (See. *Ibid.*, September 1967, p. 48).

sively as an exporting branch of economy. This is evident from the following data:[95]

Item	1947-1951	1961-1963	1964	1965
Rice output * (million t. husked rice)	1.2	1.9	2.2	2.3
Export of rice (million t.).	0.49	1.03	1.35	1.55

* Rice output in the 60's is respectively for 1961/62-1963/64, 1964/65, 1965/66.

In 15-18 years the U.S.A. has increased its rice exports approximately threefold, and now is far ahead of Burma in export volume (Burma's rice exports amounted to 1.36 million tons in 1965). And though in the first half of the 60's the U.S.A. accounted for less than a fifth of the world rice export, in addition, selling rice, thanks to its high quality, at higher prices (prices for American rice were 70-80% higher than those for Thai rice), in the nearest future the U.S.A. will be able to significantly increase its exports still further, thereby pushing down market prices on this product. Already now it is possible to observe that prices on American rice tend to fall. For instance, in New Orleans—the main rice wholesale market in the U.S.A.—prices on similar sorts of rice dropped from 21.9 dollars per 100 kg in 1950-1954 to 19.6 dollars in 1961-1965, or by about 10%.[96] In other words, the U.S.A. is stepping out more strongly as one of the biggest competitors of the developing nations—the chief exporters of rice; besides, this competitor is all the more dangerous potentially because the rapid growth of rice production and exports is accompanied in the U.S.A. by a steady reduction in the cost of production.

It is interesting to note that the world market prices on rice (and indirectly, the world prices on wheat, through the mechanism of influencing the former) have a regulating effect

---

[95] "Statistical Abstract of the United States 1965", p. 657; "Monthly Bulletin of Agricultural Economics and Statistics", December 1966, pp. 19-20.

[96] "Production Yearbook 1964", p. 339; "Monthly Bulletin of Agricultural Economics and Statistics", September 1966, p. 55.

on the domestic, national prices on rice, defining the general dynamics of their fluctuations in certain rice-importing countries. The Philippines represent a typical example (see Chart 2). It is important to note that the national price on rice in the Philippines is very sensitive to any fluctuations on the world grain market. For instance, the movement of the Philippine price practically totally coincided with the movement of world prices on Canadian wheat in 1950-1955, but it also responded sensitively to fluctuations in the movement of world (i. e., Burmese and Thai) prices on rice in 1957-1959.

Therefore, the preservation of a strong dependence of the economy of the developing nations on world capitalist economy and its market means that the process of price formation on the national food markets in a number ot instances proceeds, in a certain degree, quite independently in relation to the effect of internal economic factors which determine the cost categories (national productivity of labour, national cost of production). To a considerable extent, these processes are subjected to the regulating influence of cost categories of the developed capitalist countries. The main result of this influence, at the present stage, is the growing difficulties in agricultural reproduction in the developing countries.

It is important, however, to keep in mind that in contrast to industry, as a rule, government measures of protectionism do not cover agriculture in the developing nations. That is why the commodity production in this branch of economy is almost entirely defenceless against the oppressive influence on the part of the world capitalist market. Appearing directly in the grain exporting developing countries, this influence has an ever-increasing effect in the food importing countries, as the national food markets of the latter are gradually sucked into the general system of the world capitalist market. That is why the problem of protecting agriculture from the destructive influence of the laws of the world capitalist market objectively arises quite acutely in a number of national states.

The same problem arose in the 1950's in such a country as Japan, which inherited a very backward agriculture from the 1940's. The Japanese bourgeoisie, having put into effect an agrarian reform, is striving to stimulate the growth in the absolute volume and economic efficiency of grain production by maintaining state purchasing prices on the main crop—

rice—at a very high level. For instance, the purchasing prices for husked rice, which in 1954-1960 stood (in U.S. currency) at 18.5-19,3 dollars per 100 kg went up to 20.5 dollars in 1961, 22.6 in 1962, 24.5 in 1963, 27.8 in 1964 and 30.3 in 1965 and 1966.[97] In the course of these 13 years Japanese purchasing prices exceeded by 1.5-3 times the prices on rice of the world's main rice exporters—Burma and Thailand. And even now they are 1.5 times higher than the export prices on American rice. In other words, the Japanese government keeps purchasing prices at a level which corresponds to the individual cost of rice produced not in the best, and apparently not in the average conditions, but in the conditions lower than average. Without such a stimulant, the large farmers, and even certain groups of rank-and-file producers, would hardly be able to re-equip the technical basis of their farms on the scale they managed to do in the past decade. Only an intensive policy of state agrarian protectionism was able to stave off the fatal influence of the laws of the world capitalist market on the process of expanded reproduction in Japan's agriculture which still remains the most backward branch of the economy, the "heel of Achilles" in the country's economic growth.[98] As very aptly put by B. N. Ganguli, the distinguished Indian scholar, citing the report of the noted economist, H. Haberler, "Trends in International Trade, GATT", "while industrial countries have been pursuing generally protectionist policies to stimulate agricultural development, 'agricultural antiprotection devices are to be found mainly in some of the less industrialized countries where they tend to descourage agricultural production.' Is this not a paradoxical policy for a country which aims at agricultural development and selfsufficiency in foodgrains?"[99]

The contemporary history of some developing countries has already witnessed instances when appreciable successes have been attained in developing agriculture, particularly

---

[97] "Production Yearbook 1964", p. 339; "Monthly Bulletin of Agricultural Economics and Statistics", September 1967, p. 46.

[98] For more details about agrarian protectionism in Japan, see: "Современная японская деревня", М., 1964, гл. VI (В. А. Попов), стр. 112-116, 119 ["The Japanese Village Today", Moscow, 1964, Ch. VI (V. A. Popov)].

[99] B. N. Ganguli, Agricultural Development and the Price Mechanism, *vide*: "Indian Journal of Agricultural Economics", Bombay, vol. XIX, 1964, No 1, p. 23.

in grain farming, through a policy of agricultural protectionism.

The introduction of high state purchase prices on grain (at a level exceeding world prices) at the end of the 1940's and the beginning of the 1950's, which had a decisive effect on the dynamics of grain prices on the free national market, played, in our opinion, for practical purposes, the most important role in accelerating the growth rates in grain production in Turkey.

On the contrary, when this policy was finally given up in the mid-50's, with the pressure of the laws of the world capitalist economy on the national market being stepped up, the scope for market incentive influences was greatly restricted. And this was one of the basic reasons behind a sharp decline in the growth rates of grain output.

Of the developing countries of Asia today Ceylon is carrying out the policy of agricultural protectionism. The government maintains a guaranteed purchase price of rice, the staple cereal, at a level considerably exceeding the price of imported Burmese rice. Indeed, in 1965 and 1966 the purchase price of paddy (in rice equivalent) was $17.8 per 100 kg, while Burmese rice was imported by Ceylon at a price of $10.4 (F.O.B. prices at Rangoon). On the whole, Ceylonese producers received for their rice from 50 to 60% more than, for example, Indian rice growers. The policy of relatively high purchase prices is by far not the least of all factors responsible for the average annual growth rates of agricultural output in Ceylon exceeding those in many other developing countries of the ECAFE region: they amounted to 3.5% on the average in 1960/61-1964/65, as compared with 2.4% for the whole ECAFE region. However, while striving to expand grain production through a protectionist policy, the Ceylonese government makes definite outlays for subsidizing a substantial part of the sale price of rice from budget revenue. In countries with a strikingly pronounced class polarization of the population, such as India, an ample fund for such financing could be obtained already now—by exemption through state taxation of a bigger part, than at present, of enormous net incomes of the top strata of society (landlords, big traders and businessmen and, naturally, monopolists) and especially by complete appropriation by the state of all capital gains and speculative profits.

American food aid plays a specific role of a conductor

through which the laws of the world market exert their pressure on national economies.

In certain countries, the American wheat delivered under PL 480 really makes up a large part of the overall marketed wheat sold on the home market. In India, for instance, this share amounted to 50.1% in 1960, 32.5% in 1961, 35.9% in 1962, 47.4% in 1963, 57.5% in 1964, 53.9% in 1965 and 58.4% in 1966 (as against 10.8% in 1956—the first year India received American grain under aid programme).[100] American wheat is sold on the domestic market at a price lower than the national wholesale prices. In other words, the wheat is sold very "cheaply". But the sale of such a huge quantity of "cheap" overseas wheat on the home market must inevitably make its market price one of the most important direct regulators of market prices for the national wheat, with a tendency of "freezing" them at a relatively low level. That is precisely what happened in India. The dynamics of grain prices in that country is characterized by a fairly long period during which there is a broadening of the relative gap in price levels for the entire group of grains, and in particular, rice, on the one hand, and wheat, on the other. Chart 3, compiled according to the indices of the national wholesale grain prices in India[101] shows that the broadening of this gap actually began way back in 1956 and continued (with a break in 1959) right up until 1963. What is more, the prices for wheat were actually in a state of stagnancy. Only beginning with 1964, when the consequences of several preceding poor harvest years were felt most acutely and when an uprecedented shortage of grain arose, and the imported grain (due to the rapid development of speculative tendencies and the influence of other factors) could no longer bridge the gap in supplying the nation with food—only then did the entire system of stabilizing prices on the grain market collapse and the price of wheat jumped up sharply. As a result, the gap between the price indices for the two chief types of grain narrowed. Generally, a similar situation was observed in Pakistan (see, for example, the dynamics of wheat prices in Chart 1), with the only difference that the swift growth of national wheat prices followed, as in India, the appearance of a colossal grain deficit,

---

[100] S. C. Gupta, *Op. cit.*, pp. 38-39; "The Far East and Oceania Agricultural Data Book. May 1967", pp. 16, 60.

[101] "Bulletin on Food Statistics. February 1966", Delhi, p. 151.

Chart 3

The Dynamics of Wholesale Prices on Grain in India, 1950-1965

(1952/53=100)

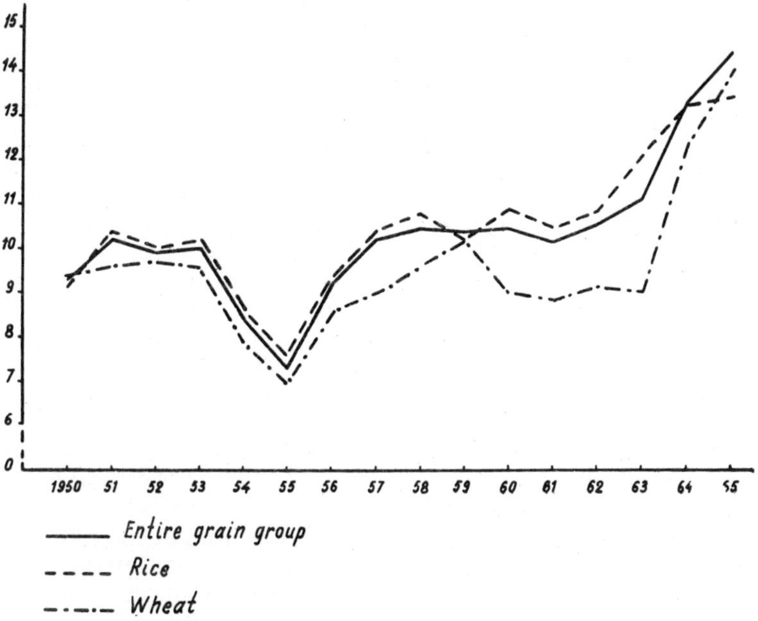

and correspondingly the sharp increase in the current demand for grain oversupply began to take place at the end of 1966.

In both cases, it is possible to observe another aspect of the above-mentioned economic law, namely: when there is insufficient supply (whatever the reason may be), the market prices on the commodities of a given branch are regulated by the value of that part of these commodities which is produced in less favourable conditions, i. e., at high (or comparatively high) costs of production, which is quite typical of agriculture in the developing countries.

On the wheat markets of India and Pakistan the increasing role of the influence on market prices of the national cost factors (by reducing the influence of international cost factors) occurred, as we have seen, in a spasmodic manner, within a couple of years. The picture is somewhat different with the rice markets of these countries.

The national rice markets in both countries are too large to be directly affected by the world rice market in any sizable manner. (In India, for example, the import of rice in the first half of the 60's accounted for only 5% of the gross marketed rice consumed in the country. Besides that, the prices on rice from one of the biggest suppliers of this product—the U.S.A.—remained much higher than the national prices for quite a long time). The national rice markets, however, were subjected to indirect depressive effect by world wheat market. As was pointed out in one of the investigations by the Reserve Bank of India, "wheat imports acted as a moderating influence on prices of other cereals by effecting a shift in demand from other cereals to wheat...".[102] But by and large, the depressive influence of the laws of world capitalist economy affected to a lesser degree the national rice markets, where the price formation proceeded more independently in relation to the price formation process on the world market; it was, in fact, more closely linked with the action of the internal economic laws.

Under the influence of huge imports of American food products and due to certain other factors that we shall omit here, the situation that came about in India was such that the commodity exchange between different sectors of the national economy for a long time was conducted to the detriment of the grain producers, especially the wheat producers who suffered the greatest losses. This is vividly seen in the dynamics of wholesale prices of the chief groups of commodities on the country's home market (see Table 17).

These figures reveal that the gap in the relative price levels for manufactured goods, on the one hand, and for wheat, on the other, was widening right up until 1963/64. This gap was especially wide in the early 1960's. For instance, the Agricultural University of Punjab which conducted an investigation into the conditions of marketing of wheat in the state established that the wheat producers received for their produce a price that in 1960/61 was 10.7%, in 1961/62 17.5%, and in 1962/63 24.3% below the "parity price".[103] It is obvious that the losses which the wheat produc-

[102] "Reserve Bank of India Bulletin", June 1967, p. 755.

[103] A. S. Kahlon, S. S. Grewal, A Study on Marketing of Wheat in Punjab, Ludhiana, 1965, p. 53.

Table 17

Indices of wholesale prices in India *
(1952/53=100)

Years	General index	Manufactures		Grain		
		Total	Finished products	All cereals	Rice	Wheat
1955/56	92.5	99.7	99.6	76	78	72
1956/57	105.3	106.3	105.6	96	97	88
1957/58	108.4	108.1	108.2	101	105	88
1958/59	112.9	108.4	108.1	107	105	105
1959/60	117.1	111.7	111.3	104	105	96
1960/61	124.9	123.9	122.8	104	108	90
1961/62	125.1	126.6	124.6	102	105	91
1962/63	127.9	128.8	127.1	106	111	90
1963/64	135.3	131.1	129.7	116	125	99
1964/65	152.7	137.3	134.8	139	134	130
1965/66	165.1	149.2	145.4	148	141	138
1966/67	191.2	163.0	156.1	174	173	158

* "Reserve Bank of India Bulletin", February 1962; *ibid.*, June 1967.

ers suffered in the course of the market exchange were very large.

This gives rise to the following question: how is India's national grain production, in particular that of wheat, affected by large injections of American grain into the Indian economy? Certain Indian economists are prone to believe that PL 480 grain imports practically do not affect (or do affect but only with a minimum effect) the economics of Indian grain production.[104] Indeed, if we take the Punjab, for instance, the following changes in the production of wheat are immediately visible: the share of wheat acreage in the overall cropped

Parity price is the price the salesman would have received if the dynamics of the prices for this product coincided with the dynamics of prices on the goods he buys. The ratio of prices in 1954/55-1956/57 was taken as the initial basis (i. e., equalling 100).

[104] "On the whole, taking the available data on acreage under cultivation, agricultural investments, and farm inputs, it seems that import of P. L. 480 foodgrains has not adversely affected agricultural production." (V. K. R. V. Rao and Dharm Narain, Foreign Aid and India's Economic Development, Bombay, etc., 1962, p. 43).

area increased from 20.8% in 1950/51-1954/55 to 22.6% in 1958/59-1962/63; the gross wheat harvest in the same period also grew by 48.6%, and on the whole, the average annual rate of growth of wheat output from 1950/51 to 1962/63 amounted to 5.1%.[105] The fact that there was an expansion in production, and a significant one at that, is quite evident.

But what factors were responsible for this growth? First of all, there was a significant increase in the acreage of irrigated farming. As a result, the wheat farming was ousting the millet one, since even at low prices it is more profitable to grow wheat on irrigated land than it is to grow millet. It is not surprising, therefore, that the main increase in wheat production was registered in the southwest regions of the Punjab (including the state of Hariana which was formed in 1966), where the Bhakra-Nangal Canal flows. Secondly, a large part of the increase in wheat production came from the petty-commodity structure, i. e., subsistence farms, where the growth in food output takes place to a considerable extent irrespective of the prices, since the main purpose of production here is, first of all, to increase the consumption of the producer and his family. The tendency to reduce the level of marketability of wheat farming, registered in the Punjab, in spite of the formation of capitalistically-developed regions (see Section III) with a generally noticeable inrcease in the wheat output in the state, is apparently ample evidence of the fact that the petty producer played a very important role in expanding the wheat production in the Punjab.

On the whole, the "freezing" of market prices on wheat for a comparatively long period of time had a retarding effect on the general growth of production of this second most important grain crop on the Indian subcontinent, since large groups of rural producers (first of all, specializing in production for the market), in such circumstances, were deprived of their main economic stimulus for expanding production. Moreover, the small accumulation fund which ultimately accrued to the productive sections of the agricultural population had a tendency to shift into the production of industrial and other crops for the marketing of which the situation was more favourable.[106] Grain farming remained, as hitherto, the function, in

---

[105] A. S. Kahlon, S. S. Grewal, *Op. cit.*, pp. 2, 3, 6.

[106] Referring to studies of trends in yields of rice and wheat in India (revealing that wheat yields remain stable while rice yields increased),

the main, of low productive economic structures with their pivotal link—the farm of subsistence type.[107] However, as was so justly pointed out by S. C. Gupta, quoted earlier, "development measures during these years have more than compensated for the disincentives provided by the price factor. If both factors would have operated in a similar and not in contrary directions, the results could be expected to be much better."[108]

The fact that imported grain coming in through governmental channels is sold on the home market not at the spontaneously prevailing national prices, but at prices approximating the world prices, is explained by a number of reasons, the most important of them being the desire of the ruling classes to stabilize the food situation in order to ward off any aggravation of social conflicts which are inevitable when there is a shortage of food, blackmarketeering, etc. The real chance for a policy of such stabilization, falling back on imported grain, lies in relatively low world price on wheat. That is precisely why the ruling classes in India and Pakistan can afford the sale of colossal amounts of imported grain on the home market at "cheap" prices: for instance, the wholesale issue price on American wheat in Dacca (Pakistan) in 1962-1965 was equal to 6.2-7.9 dollars per 100 kg; in India from the late 50's to 1964 it stood at 7.9 dollars per 100 kg (in 1965—10.1 dollars), while similar sorts of wheat sold on the wholesale market in Kansas City in the U.S.A. at prices which varied from 1958 to 1963 within a range of 7.1-7.5 dollars per 100 kg (in 1962—8.3 dollars).

If the world price on wheat were double today's price, then for instance, in India, with the 1966 volume of wheat imports

---

the distinguished Indian sociologist, Prof. K. N. Raj notes: "At least part of the explanation lies in the fluctuations of wheat prices caused by large releases of imported wheat creating more uncertainty than for rice and thus persuading growers to apply inputs like fertilizers to other more dependable crops like cotton where there was such a choice". (K. N. Raj, Indian Economic Growth. Performance & Prospects, New Delhi, etc., 1965, p. 5).

[107] In the mid-1960's the market situation in India has essentially changed in favour of agriculture. The main benefits, however, as shown by various surveys, are reaped not by the mass of agricultural producers, but by economically dominating groups, who monopolize the agricultural markets (for more details see the next section).

[108] S. C. G u p t a, *Op. cit.*, p. 45. For details about the effect of American food aid on India's agriculture, see also: R. A. Ulyanovsky, The Dollar and Asia. U. S. Neo-Colonialist Policy in Action, Moscow, 1963.

Sold at the 1964 wholesale prices on the home market, the state, in order to cover the difference between the world price and the home one, would have to spend the tremendous sum of approximately 700 million dollars (or, according to the former rate of exchange for the rupee, more than 3.3 thousand million rupees). Naturally, this sum would have to be taken out of the national savings fund. But even with the present-day level of world prices, the policy of stabilization calls for certain material sacrifices on the national economy, running into 20-50 million dollars (90-230 million rupees) annually[109] in the second half of the 50's and early in the 60's (defrayal of part of the transportation expenses, certain subsidies for the selling price on grain received under the aid programme, and some other expenses).

The policy of stabilizing grain prices with the backing of imported grain has a definite influence on the re-distribution of the national income. Since this policy is implemented at a time when there is a general rise in nominal commodity prices, by keeping the nominal wages of the working people from rising, the capitalist class gets significant additional opportunities for accelerating the accumulation of capital. However, by its negative influence on market grain prices (in the first place, on wheat prices), it worsens the conditions for reproduction on farms, primarily of a petty commodity structure, which must sell their marketed grain at a price below its cost. In other words, such a policy of stabilization, on the whole, is implemented at the expense of the productive forces in agriculture (first of all, in grain farming),[110] since it leads to a shift—without subsequent reimbursement—of part of the income created in rural areas to the urban communities, primarily those of large-scale capitalistic structure. In this respect, such a policy tends to enhance the trend so typical of many developing countries—that of a growing non-equiva-

---

109 S. C. Gupta, *Op. cit.*, p. 40.

110 In particular as was pointed out in a comprehensive report on the world food problem, prepared by a group of American experts, "food aid may influence the price behavior of agricultural products in the recipient countries. The governments of these countries, under strong political pressures from urban consuming groups, may be tempted to use food aid to keep market prices for food below the level required to provide adequate incentives for their own agricultural producers" ("The World Food Problem. A Report of the President's Science Advisory Committee", vol. II, Washington, 1967, p. 183).

lent exchange between different structures within the framework of the national economy.

The policy alternative to stabilizing market prices, backed by imported grain, is one of general stabilization of prices on national agricultural produce and manufactured goods, combined with certain social measures that ensure this variant of stabilization (this will be dealt with in the following section).

Thus, the growing gap in the economic levels of development between the industrially advanced capitalist countries, on the one hand, and the emerging countries, on the other, the close links of the latter with the world capitalist economy, give rise to a number of dovetailing problems: world capitalist economy, by expanding the sphere of influence of its economic laws, retards productive savings for and worsens conditions of reproduction in agriculture, depriving the national economy of more opportunities for reproducing the national food fund within the necessary limits. Moreover, the depressive influence of the economic laws of world capitalist economy in certain cases (in certain regions, for certain types of crops), can attain such proportions that the expansion of grain production (if such occurs) will not result from developing commercial, enterprising farms, but will rather be the consequence of increasing the number of petty, actually subsistence farms, which are able to withstand such pressure only by selling their "surplus" products at a price incommensurate with the cost of production.

On the other hand, in view of the growing shortage of food resources, the developing nations are compelled to replenish them, to an ever-increasing degree, from the vital "surpluses" existing in highly-developed capitalist countries, thereby turning their food markets into an integral part of the world capitalist market, and in so doing, exposing new sectors of their national economy to the influence of the economic laws of world capitalist economy. In other words, with a model of socio-economic development that is upheld by the ruling classes of many developing countries, the national yearning to break the semi-colonial structure of economy is counterposed by the tendency to intensify the dependence of national reproduction on the reproduction in the developed capitalist countries. Moreover, reproduction of national food resources becomes an organic (and dependent) element of agricultural reproduction in the developed capitalist countries.

THE MARKET AND REPRODUCTION IN GRAIN FARMING. THE INFLUENCE OF NATIONAL FACTORS

So far we have examined the influence of the mechanism of the world market on the price formation process in the domestic market and those negative consequences which stem from this influence on food production in the developing countries. This section deals with certain internal factors defermining the price formation process, factors which also retard the growth of the productive forces in agriculture.

The arrivals of marketed grain at the local markets are extremely uneven throughout the year. The arrivals are especially great in a comparatively short period—the period following the completion of the production cycle (harvesting, threshing, etc.). For instance, in India the bulk of marketed wheat reaches the market in the course of only four months from April to June (in 1964 this period accounted for 61% and in 1965—64% of all wheat arrivals), the bulk of sorghum (jowar) arrived also in four months: December-March (in these months in 1964 59% of all the sorghum deliveries were realized in the market and in 1965—61%).[111]

This mass flow of grain to the market within a comparatively short period is explained, first of all, by the fact that the petty commodity producers, who are very important suppliers of marketed produce, are in dire need of money and strive to sell their marketed "surplus" as soon as the crop is ready for sale. The credits which these producers receive do not prevent them from flooding the market with their grain; on the contrary, they act as a factor compelling them to sell their produce immediately (because, as we have seen, the credits are given them at high interest ratés, quite often for a period up to the harvest time; the creditors not infrequently buy up the crops still in the fields at a price previously agreed upon, etc.). Many groups of petty capitalist farmers are also forced to immediately throw large consignments of grain on the market because, the capitalist credit system being poorly developed, they are in dire need of ready cash to carry on reproduction on their farms. In other words, the contingent of people immediately selling their produce on the market is quite broad and diversified.

The irregular dumping of produce on the market, giving rise to a temporary situation when supply on an increasing

---

111 "Agricultural Situation in India", August 1966, p. 397.

scale exceeds demand, on the one hand, the monopoly on the market of trading and money-lending capital taking full advantage of such a situation, on the other, trigger very significant seasonal fluctuations of market prices. A typical feature of these fluctuations is that prices drop sharply when large quantities of grain are put out on the market, and, on the contrary, prices jump up (compared to the annual average level) when there is a drastic cut in deliveries to the market. The range of price fluctuations can be seen from the situation on the grain markets of India and the Philippines (see Table 18).

Table 18

India (the Punjab) and the Philippines: seasonal fluctuations of prices on grain markets *

Month	India (the Punjab) Wheat		Month	Philippines, Palay—unhusked rice	
	market arrivals % (aver. 1954-1963)	seasonal price index (average 1954-1963)		produce harvested, % (1963/64)	seasonal price index (average 1949-1958)
April	4.0	99.1	July	2.2	105.2
May	22.9	92.2	August	4.8	104.9
June	30.2	94.4	September	5.9	107.9
July	10.0	94.9	October	15.3	107.7
August	5.7	96.4	November	19.2	105.7
September	4.7	96.7	December	17.7	94.5
October	4.8	99.4	January	14.2	90.9
November	4.0	102.3	February	5.2	93.2
December	4.1	104.6	March	4.7	92.8
January	3.2	107.4	April	7.3	96.6
February	2.9	108.1	May	1.2	99.2
March	3.5	104.5	June	2.3	101.4

* A. S. Kahlon, S. S. Grewal, *Op. cit.*, pp. 34, 37; J. Gutierrez, Rice Marketing in the Philippines, *vide*: "Economic Research Journal", vol. XII, 1965, No. 3, p. 214.

There is a sharp drop in seasonal prices in the Punjab in May when the market is flooded with large quantities of grain; moreover, the period of low prices in the state lasts for three months, in the course of which 64% of all the marketed wheat arrives at the market. When the mass arrivals stop—beginning with August—prices tend to grow, reaching a peak in January and February, i. e., shortly before the next crop is ready for harvesting. In the Philippines the sharpest drop in prices is observed in January when the market is flooded with the bulk of the paddy harvested in October (threshing and drying take two months). And again the period of especially low prices lasts three months during which period most of the annually marketed paddy supply is thrown on the market. The prices reach a peak in September-October, on the eve of the harvesting of the next rice crop. "The irrationality in the seasonal disposal of palay by farmers in the country, notes J. Gutierrez, seems to arise from the pressures of debt repayment and the need of cash... Financial pressures tied with marketing of palay make it difficult for farmers to achieve a more rational disposal of farm produce. Once debt is incurred, farmers are forced to release their palay during months of low prices. The same palay could have channelled back to these farmers at higher prices in the form of credit".[112]

The social essence of these seasonal fluctuations of market prices boils down to the following: the lowest prices are received by the main mass of producers, above all, by almost all their groups representing the petty commodity structures in the rural areas.

Meanwhile, the difference between the minimum and maximum prices is pocketed, first of all, by the merchants having a monopoly on the market and controlling the bulk of arrivals to the market. For instance, in 1960-1963 at one of the biggest grain markets in the Punjab-Jagraon—the wheat arrivals in May-July made up 65% of the yearly arrivals, while the grain offtakes by merchants for subsequent redistribution among the consumers through trade channels in these months amounted to only 35%. The bulk of the wheat (49%) was sold by the merchants in the period of rising and maximum prices (from August to April), and 16% was not sold at all,[113] and was apparently hoarded in anticipation of a new price increase.

Market fluctuations in India, in recent years marked by the predominant rise in prices on food products, in particular, on grain (see Table 17), were apparently only insignificantly favourable for the petty producers, even though they could somewhat increase their own consumption of grain products.

---

[112] J. Gutierrez, *Op. cit.*, pp. 214-215.
[113] A. S. Kahlon, S. S. Grewal, *Op. cit.*, p. 9.

The main benefits stemming from such changes are enjoyed precisely by those social groups in rural and urban communities who take possession of the bulk of the marketed produce (rich landowners, merchants, etc.). For instance, at the jagraon grain market the net profit pocketed by the merchants only as the difference between seasonal prices (price received minus price paid minus storage cost) took the following pattern (in %):[114]

Period of sale *	1953-1962 (average)	1963/64	1964/65
November-December	4.1	15.5	29.8
January-February . .	6.9	35.1	27.8

* Period of purchase: May-June.

In the years when there was a sharp rise in grain prices, the norm of net profits of the merchants, on approximately six-to-eight-month turnover of capital, grew 4-8 times and exceeded by 1.5-2 times the norms of profits of big industrial companies. (In the light of sharp market fluctuations in the country it is possible to understand the recent feverish efforts of the Indian monopolies to gain a foothold—land, etc.—in agriculture, which they have so far ignored as an object for capital investments.)

Therefore, even if we take only internal factors influencing price formation on grain markets, completely disregarding international factors, for the time being, it is clear that the market mechanism, controlled as it is by forces interested only in maximum profits and exploiting agricultural production, acts to the detriment and against the best interests of the main mass of grain producers. Through the sphere of the market and commodity exchange and due to the simple "rigging of prices", outwardly spontaneous but in reality guided, this mass of producers are deprived of a considerable slice of their potential savings. And all this is in addition to the losses they must suffer due to payment of rents and interest to landlords and money-lenders. From all this, it is quite obvious how difficult it is to expand reproduction in grain farming.

[114] Calculated on the basis of: *Ibid.*, pp. 18, 27.

As a result, grain farming, which is the most extensive branch in agricultural economics, has in many countries come to a "standstill" primarily in the subsistence stage (excluding, of course, certain groups of farms and even entire regions where large-scale, commercial production has a good foothold). In many instances, it becomes economically unprofitable to conduct such farming. The Filipino economist quoted earlier, D. A. Maulit, gave the following appraisal of the economics of rice farming in the Philippines, "Rice production in this country is more for home consumption than for the market. The possible reason is that rice has not yet become as good a source of cash income as some other crops... There is no sufficient incentive to increased production in the rice industry."[115] It is precisely for this reason that in many developing nations there exists a great gap in the rates of growth of technical and other crops, on the one hand, and the grain crops, on the other hand.[116]

It should be noted that in certain countries, at certain stages of their history, the government tried to retard the changes in the structure of agriculture, and avert the process of a relative reduction in food crop areas, mainly under grains. (The case in point is, for instance, the "Grow More Food Campaign" in India, which included, among other things, such measures as increasing acreage sown to grain by reducing the areas planted to technical crops. Since these attempts affected the existing production of technical crops, they clashed with the law of value, whose play in the existing circumstances was reflected in the substitution of "profitable" crops for "non-profitable" crops. But such a policy could not but fail precisely because it was not backed up by all-out measures to remove the obstacles blocking the development of commodity production in agriculture.

---

115 D. A. Maulit, Statistics on Disposition of Philippine Palay Production,—"Economic Research Journal", vol. III, 1957, No. 4, p. 210.

116 For instance, in many Asian and North African developing countries, the rate of growth of production of raw farm produce for the processing industry is approximately 1.5-2 times higher than the rate of growth of food grain production. In India, for example, for the 50's and early 60's along with an approximate 40% increase in gross agricultural output, and a 35% increase in grain and pulses harvests, the output of fibre crops increased by 75-100%, sugar cane harvests—by 60-70%. In Pakistan where from 1947/48-1951/52 to 1960/61-1962/63 grain harvests increased by only 22%, the volume of cotton fibre produced increased by 51%, sugar cane—by almost 100%.

Without fear of exaggeration, one may say that the process of transforming grain farming from a predominantly subsistence economy into a paying proposition will represent a whole epoch in the economic evolution in the countryside. One can hardly expect any noticeable effect from capital investments injected into agriculture with a view to raise food crop output, unless a simultaneous attempt is made to re-establish economic stimulants promoting the growth of production for commercial purposes.

One of the most important stimulants is, of course, such market prices as would boost production.

In the social situation existing in many developing nations the problem of regulating prices on food products, and first of all, on grain, is one of the most complicated ones. A noted Filipino statistician, I. C. Belarmino, puts it this way, "The price dilemma of rice is a two-edged sword. While consumers complain about the high price of rice at the corner store, producers lament the very low prices of palay received by them on the farm. Consumers clamor for importation and the establishment of a price-ceiling; the producers, for no-importation and price-support. What steps, then, should the government take?"[117]

In such an approach to the question, serious thought should be given to the broad and well-founded recommendations of the eminent Indian economist, Prof. D. R. Gadgil, on the problems of agricultural price policy.

In his recommendations, Prof. Gadgil proceeds from two main principles:

1. Prices on all types of agricultural products must be stabilized; moreover, the fixing of prices received by the producer is based on the actual cost of production (hence—the "remunerative" or "profitable" price), while the price to be paid by the consumer is based on the price received by the producers, the actual cost of delivery and distribution plus the small profit, strictly controlled, for the marketing agent (private trader or cooperative or government organization). In other words, there should be a stable structure of agricultural prices among which definite ratios are established. And likewise, there should be a definite ratio between prices for industrial goods and prices for agricultural products.

---

[117] I. C. Belarmino, The Rice Problem: An Appraisal, *vide*: "Economic Research Journal", vol. XII, 1965, No. 1, p. 4.

2. Since the implementation of these measures is out of the question in conditions of a spontaneous private market, strict state control must be established over the market, moreover, the state should set up a special mechanism for carrying out this control, for purchasing and distributing farm produce at fixed prices.[118]

As for grain farming, the introduction of stable prices on grain, bearing in mind the actual cost of production, would offset the losses suffered by the producers due to the low prices during periods of mass disposal of grain on the market; and on the contrary, the profits of the traders would be reduced to a set level. This fact alone would considerably improve the conditions for reproduction in grain farming (since the savings fund would increase) and would give a definite impetus to enlarging acreage planted to grain. Of course, as a result of larger receipts for one and the same amount of products, marketed grain deliveries from the petty commodity structure could fall, since the petty producers would increase their own consumption (this tendency, as was already mentioned, is of universal nature). But on the other hand, in terms of the market conditions, the grain crops would be placed at par with technical crops and would become attractive for any farm producing all or part of their products for commercial purposes.[119] Another equally important aspect of stabi-

---

[118] At a seminar of Indian economists on problems of planning, held in Delhi in December 1966, Prof. Gadgil worded his stand on the policy of agricultural prices in the following way: "...any plan of increased agricultural production in specific directions cannot succeed without the maintenance of a relative stable structure of prices of different crops. The failure of our food policy, in the recent past, was due, in part, to allowing prices of some foodgrains to reach unduly low levels. Also... in view of the technically weak position of the agricultural producer vis-à-vis the trader and processor, stabilization of prices of agricultural products is a necessary measure for the stabilization of income of the agriculturists. Therefore, it is necessary to devise a system of purchase and/or procurement, storage, and sale and/or distribution of all the major agricultural products of the country through which adequate supplies of these products required for national policy purposes would be acquired and the prices of the products stabilized." (D. R. Gadgil, Indian Planning (mimeo), December 1966, pp. 21-22; see also: D. R. Gadgil, Price Policy for Foodgrains,—"Economic Weekly", Bombay, 1964, Sept. 26, pp. 1561-1568; D. R. Gadgil, National Economic Policy after Devaluation,—"Mainstream", Delhi, 1966, October 8, pp. 8-13).

[119] It is noteworthy that India is engaged in a broad discussion of the following question: precisely from what type of farm or group of farms can the cost of production be taken as the initial basis for

lizing grain prices is that, in his turn, the consumer buys grain on the market at stable prices which will be below those jacked up by the traders and blackmarketeers on the grain market today.

Therefore, the recommendations of Prof. Gadgil envisage a system of regulating market prices which is in accord with the demands of the law of value—the basic law of commodity production.

The problem is that to be implemented, such a policy opening up broad avenues for economic stimulants presupposes, as one of its initial prerequisites, a significant encroachment upon the positions of the economically dominating class layers, and first of all, rich landowners and traders monopolizing the market.

In cases when the ruling class is reluctant or simply unable to put into effect the necessary reforms, we actually do encounter "Kingdoms of Crooked Mirrors". Take, for instance, the Philippines. Of the commercial grain (from the 1963 harvest) that arrived at the market in 1963/64, the Rice and Corn Administration procured approximately 9%. Moreover, 82% of all the purchases were made not when the petty producers disposed of their produce (i. e., from January to March), but at a time when considerable quantities of rice were concentrated in the hands of traders and merchants, and market prices had reached a maximum level (i. e., from July to November, the months when the seasonal price index varies within 105-108).[120] The state, through its procurement agency, actually ensured tremendous profits to trading-cum-speculative capital thus depriving the petty producer of the only opportunity to receive a "fair" price.[121]

---

establishing a guaranteed price for the producer? In spite of the wide range of opinions on this account, it is interesting to note the views of those scholars who believe that "the guaranteed price should cover the cost of cultivation of the bulk of the more efficient production." Assuming that the "bulk of production" equals 85% of all the production, then in the state of West Bengal the guaranteed price would be received by 73.8% of all the farms which have control of 80.2% of all the acreage planted to Aman (winter) rice (see: M. L. Dantwala, Principles and Problems of Agricultural Price Determination,—"Technical Address. 19th Annual Conference of Indian Society of Agricultural Statistics", 1966, p. 10).

[120] See: J. Gutierrez, *Op. cit.*

[121] Just how powerful the trading-cum-speculative capital is is evident from the example of India. When the Punjab government in January of 1967 tried to transfer the burden of incidental charges, thus far paid

In certain developing nations, the government has practically unlimited control over the local markets, thereby steering the entire process of price formation for the main types of agricultural products. In those cases, however, when the state policy of agricultural prices is not in keeping with the law of value, it may have an adverse effect on the development of production.

In this respect, it is interesting to follow the processes taking place in Burma where the existing system of price fixing gives rise to high land rents which are appropriated by the state.

State land rent is directly realized as a difference between the world price on Burmese rice (Rangoon export price) and the fixed internal purchasing price established by the state, which, beginning with 1948, concentrated practically all the marketed rice designated for export in its own hands. The stable purchasing price for rice, introduced also in 1948, remained almost unchanged until 1960; from 1961 it was increased somewhat, but not more than by 5-10%.

The portion of the world price on rice appropriated by the state is very large. In the period from 1948 to 1957, rice sellers (in the main, all kinds of intermediaries are money-lenders rather than rice producers) received for the marketed grain submitted for export (at least a third of the gross national harvest) only 43.3% of the sum which the state received from sales on the world market. In 1961 for one unit of rice exported they received 53.5%, in 1962—53,2%, in 1963—50.4%, in 1964—48.2% of the world price on national rice.[122] According to an assumed estimate, the costs of delivery of rice from local centres to Rangoon, of its cleaning and storage (for half-a-year) took another 10-15% of its world price. Therefore, the net profit cleared by the state varied from one-third to two-fifths of the world price on the national rice.

---

by the land-tillers to the grain merchants, the latter called a strike for "an indefinite period" at all grain markets in the Punjab. The markets completely stopped all operations, threatening to disrupt marketed grain supplies to the population. Nine days later the government was forced to give in. The merchants did not budge a single paisa of their swelling speculative profits (see: "The Economic Times", 1967, January 3; "Tribune", Ambala, 1967, January 9, 10, 11).

[122] According to: "Food and Agricultural Price Policies in Asia and the Far East", Bangkok, 1958, pp. 20, 21; "Production Yearbook 1964", p. 338; "International Financial Statistics" (Supplement to 1963/64 issues) p. x.; *Ibid.*, November 1965, p. 27.

However, as we have already seen, the national value of Burmese rice was unusually high due to the high manual labour expenditure. It is considerably higher than the international value regulating the world market prices on grain. That is why the price for which Burma (as well as other grain-exporting developing nations) sells its rice on the world market falls below its national value. And in such circumstances, the state, through its own system of prices, appropriates a large slice of the world price on national rice, while in a number of cases the rice producers are unable to defray their production expenses.[123]

The land rent arising as a difference in prices and appropriated by the state is the result of the monopoly.[124] This is a monopoly-type land rent the opportunity for continuous appropriation of which stems from the still supreme state ownership of land as well as from the state monopoly on the ultimate domestic markets of agricultural produce. The stagnancy of rice production in Burma in the 50's, when it was below the pre-war level, was caused by this pressure from the state monopoly land rent, as well as by pre-capitalistic forms of oppression (usury, etc.). The upsurge in rice production attained by the mid-60's was brought forth by the policy of eliminating pre-capitalistic forms ot oppression, but not by relaxing pressure from the state land rent. This policy enabled the direct producer to receive the part of the monopolistic low purchasing prices of his product which had previously been pocketed by the landlords, money-lenders and trade intermediaries. In so doing, this policy opened a kind of tap for economic stimulants of production to come into play.

Undoubtedly, a certain part of the huge land rent accumulated by the state is used for expanding the reproduction of

[123] Compare the following: in Burma the internal purchasing price of paddy (in equivalent to husked rice) in the first half of the 60's was about 3 times below the wholesale price of rice in the Philippines, where the productivity of labour was somewhat higher than in Burma, and where, nevertheless, the producers strove to get price-support on rice. And even this purchasing price was 4 times below the wholesale price for rice in the U.S.A.

[124] A similar land rent was established in South Korea beginning with 1940. See: И. С. Казакевич, «Аграрный вопрос в Южной Корее», М., 1964, стр. 142—145. (I. S. Kazakevich, "The Agrarian Question in South Korea", Moscow, 1964.) The difference between Burma and South Korea lies in that in the latter the land rent is based on the difference between the compulsorily introduced low purchasing price and the price at which the state sells the products on the domestic market.

the economy (such was the case already at the time of bourgeois rule, though today considerably more of the receipts from the land rent are used for national construction). But still another aspect should not be overlooked. The heavy burden of state land rent on commodity production in crop farming holds back the development of the productive forces in petty commodity and petty capitalistic structures which are predominant in Burma's agriculture today, thereby limiting the base for national savings. At the same time, state land rent on such a large scale could, in our opinion, obstruct the policy for creating big productive farms which, according to their economic nature, require the appropriation of part of the surplus product in order to expand reproduction.

State purchasing prices for paddy were raised somewhat in 1967; the farmers were given the opportunity to sell surplus paddy on the free market (at prices higher than the purchasing prices); all these important measures broaden the field of activity of the economic stimulants in rice farming and will doubtless promote, to a certain degree, greater savings in peasant economy.

REORGANIZING AGRICULTURE: A STRATEGIC APPROACH

> "What must be done to raise this prostrate giant to his feet?"
>
> (*H. Ayrout, Fellahs d'Egypte, 1952*)

One of the most generally acknowledged theses in modern Western sociology is that the only way to avert the oncoming threat of famine is to immediately concentrate the efforts of mankind on reducing its own population, first of all, by cutting the rate of natural population increase.

One of the typical representatives of this school of thought, P. Ehrlich (Stanford University, California, U.S.A.), proclaims: "Runaway human population growth is, in many ways, analogous to the runaway growth of cell population which we call cancer... The disease is now so far advanced that only with radical surgery does the patient have any chance of survival". Without such surgery "sometime between 1970 and 1985 the world will undergo vast famines—hundreds of millions of people are going to starve to death".[125] P. Ehrlich points to India as one of the first countries to be struck by famine....

Of course, in the present conditions, it would be rash to ignore the formidable role which birth control, especially in the emerging countries with their highest rates of population growth, could play in averting the approaching menace. (Such a policy is already being practised in many of these countries, though without noticeable success). The significance of this factor is determined already by the fact that within the existing economic framework of the developing countries, dominated by low productive structures, the natural increase of the population, in a number of instances, tends to supersede the growth of the production employment potential. That is why the ratio between the productive and non-productive population becomes even more unfavourable. However, another aspect of this concept is worth noting: present-day neo-Malthusians uphold the idea of maintaining such a "population-food resources" balance in which the "dependent variable" will be the size of the population and not production providing the food resources. Actually, they do not even remotely consider the possibility of transforming agricultural production in the developing countries with their predominantly low productive structures so as to boost output, or else they have no faith in the possibility of such a transformation. However, the concentration of a nation's efforts on production problems is the very central link that should be closed, for the entire chain of food supply to be locked.

The policy of regulating population growth, no matter how important it may be, cannot in itself change either the system of production or the current semi-starvation consumption level of hundreds of millions of people.

First of all, it is necessary to have a clear-cut understanding of the tasks which the emerging nations face in the field of increasing food production. For example, in order to stop importing grain and retain and somewhat increase the standard of per capita cereal consumption by 1970-1971, a country like India must grow well over 100 million tons of grain. If we are to take the mean annual grain production (including

---

[125] P. Ehrlich, Paying the Piper, *vide*: "New Scientist", London, 1967, December 14, pp. 652, 655.

pulses) in the period of the Third Five-Year Plan (1961/62-1965/66), i. e., 80.5 million tons, as the initial basis for the growth of production, the required rates of growth will look like this:[126]

Source of estimate	Demand in grain output for 1970/71, million tons	Necessary rate of increase, annual compound percentage
Fourth Five-Year Plan (Draft) . .	120.0	8.3
Estimate by Perspective Planning Division, Planning Commission	116.2	7.6
Estimate by National Council of Applied Economic Research . .	113.05	7.05

Undoubtedly, this is a very high rate of increase which was practically unknown to the developing nations in the past 10-15 years. In this respect, it should be noted that even in the most favourable (in terms of rate of increase) period (1949/50-1960/61), the grain output in India grew on the average by only 2.7% annually.

Consequently, the average annual rate of increase has to be trebled to secure a food supply covering minimal per capita consumption. (According to an official estimate, a food-grain output of 95 million tons is expected in 1967/68, i. e. a 17%-increase above the annual average in the Third Five-Year Plan period.) ECAFE assumes that an annual rate of increase in food output of about 5% has to be attained, if the developing countries of Southeast Asia and the Far East are to regain in the near future merely the pre-war level of the per capita food production.[127]

The solution of this more than modest task will not, of course, mean that the critical point of food consumption will have been overcome. Hence, it is only natural to ask: what must the perspective rates of growth of food production be in the emerging nations? Calculations based on the data

[126] "Fourth Five-Year Plan. A Draft Outline", Delhi, 1966, p. 184; "Draft Fourth Plan, Material and Financial Balances 1964-65, 1970-71 and 1975-76. September 1966", Delhi, p. 40; "Long-Term Projections of Demand for and Supply of Selected Agricultural Commodities 1960-61 to 1975-76", New Delhi, 1962, p. 91.

[127] "Economic Survey of Asia and the Far East 1966", p. 108.

of FAO[128] indicate that, in order for the developing nations in the region of Southeast Asia and the Far East, now experiencing the most acute deficit in food products, to be able to fully stop importing grain from the world market by 1985 and to provide for its increased population (1,446 million against 896 million in 1965), the per capita food consumption of 2,600 calories per day, i. e., 17% more than the FAO's minimum standard estimated for this region (2,230 calories), then even with the development of other spheres of food production (it is assumed that grain production in these countries, as in the 50's and 60's, will account for approximately 70% of all calories), they will have to harvest about 304 million tons of grain (5 main kinds), against 144 million tons in 1962-1965. In such conditions, the average annual rate of increase for grain production must be at least 3.8%. Every tenth per cent of this increase signifies tremendous absolute figures. For instance, a drop in the average annual increase to 3.3%, which, however, is 1.5 times above the level typical of the first half of the 60's, will form a grain deficit in the developing countries of that region amounting to 29 million tons by 1985.

However, the emerging nations are in need of not simply considerable increase of food, but an increase in which the marketed mass would grow at a faster rate, i. e., to a point when the marketed food supply would correspond to the actual demand. A preliminary estimate indicates that by 1985 the demand for marketed grain in the developing Asian countries could amount to more than 50% of its gross output (whereas today grain farmers supply only 25-30% of their output to the market).

It is important to note that in a developing economy, where the process of diversification of national production is taking place at faster rates and its specialization is getting even more pronounced, it becomes ever more difficult to resolve this dual problem simply on the basis of an extensive development of semi-subsistent farming. The entire course of economic development confronts the developing nations, with mounting imperativeness, with the task of systematically raising the productivity of labour in agriculture, which has

---

[128] "Monthly Bulletin of Agricultural Economics and Statistics", September 1967, pp. 11, 14; "The State of Food and Agriculture 1967", p. 157.

gained the status of the most important factor in ensuring the national food fund in necessary quantities.

Due to the extremely low level of the productivity of labour in the developing nations, a large amount of the food produce enters the consumption fund of the producer, his family and the hired workers involved in the production process. This concerns both the pre-capitalistic and petty commodity structures, as well as the petty capitalistic structure for which an old technical basis is quite typical. (In the ideal case, with semi-feudal or state pressure eliminated, not 70-75% but considerably more of the grain produce would be consumed in the sphere of agriculture). That is why agriculture cannot supply spheres having no connection with the output of food produce (industry, etc.) with the amount of food necessary for the reproduction of manpower engaged in these spheres. On the other hand, the raising of labour productivity in agriculture by equipping it with as many means of production as possible, the development of irrigation, etc., will lead to an increase in the productivity of land, and correspondingly, to the formation of a greater market surplus which is so badly needed to provide the above-mentioned spheres of the economy with a sufficient quantity of food-stuffs. That is why the transformation of poor productive, but predominant (or in some cases, singularly dominant) structures in agriculture has been and remains a pressing necessity for the developing nations. And one of the prerequisites for such a transformation is the elimination of the pre-capitalistic forms of land ownership and capital, which in the given structures play a decisive role.

Radical agrarian reforms—and herein lies their vital economic significance—will ensure the full tapping of the large potentialities for a growth in production inherent in individual peasant farming. For example, the agriculture of the Chinese People's Republic, after carrying out a land reform throughout the country (1950 to 1952) in the next three years (1953-1955) achieved quite a high average annual rate of growth in the cereal production —5.9%. (It should be taken into account that even mutual assistance groups encompassed approximately only half of the peasant households and cooperatives of the lower type—14%)[129] This pace, how-

[129] Jean-Emile Vidal, Où va la Chine? Editions Sociales, Paris, 1967,

ever, has been completely lost since the period of the "big leap". The realization of all the possibilities for developing society's productive forces on the basis of peasant land ownership is objectively one of the most important tasks at the present stage of the socio-economic process in many emerging nations.

Even though a free development of peasant farming, based on comprehensive state aid, could help boost agricultural production, it goes without saying, however, that the long-range solution of the food problem lies in a gradual development of a system of large farms having a high level of labour productivity and marketable output. And this is the pivot and core of the whole production problem.[130]

In countries like India, Pakistan or the Philippines, the ruling circles strive to place the main stake on developing large-scale farming by gradually transforming the rich semi-feudal landowner into a large producer of capitalist type and by consolidating the economic positions of the upper layers of the peasantry. The high concentration of land, this basic factor in agricultural production, in the hands of big semi-feudal landowners, though it was shaken, in the course of land reforms,[131] and in some countries quite significantly, still has not been eradicated. In West Pakistan, for

---

[130] One cannot but agree with the following conclusion made by American experts: "The transition from traditional farming to modern agriculture will be difficult and expensive for the hungry nations but it is absolutely essential if their food needs are to be met. There is no alternative" ("The World Food Problem. A Report of the President's Science Advisory Committee", vol. I, p. 19). The root of the problem is on what socio-economic lines this transition is to be brought about.

[131] The problems of agrarian reforms have been dealt with quite extensively both in Soviet and foreign literature.

Besides a number of monographs on individual countries, several joint-effort works on this topic were published in the USSR, for instance: "Аграрные реформы в странах Востока", М., 1961 ("Agrarian Reforms in Countries of the East", Moscow, 1961); «Аграрно-крестьянский вопрос на современном этапе национально-освободительного движения в странах Азии, Африки и Латинской Америки», М., 1965 ("The Agrarian-Peasant Question at the Present Stage of the National Liberation Movement in Asian, African and Latin American Countries", M., 1965); «Аграрные реформы в развивающихся странах и странах высокоразвитого капитализма», М., 1965 ("Agrarian Reforms in the Developing and Highly Developed Capitalist Countries", Moscow, 1965); «Социально-экономические последствия аграрных реформ и социальная структура деревни в развивающихся странах Азии и Африки», М., 1966 ("Socio-Economic Consequences of Agrarian Reforms and the Social Structure in Rural Areas of Developing Asian and African Countries", Moscow, 1966).

instance, where the landlord class controlled up to half o all the land area, the last reform (1959) had stripped the rich semi-feudal land owners (having over 40 hectares of land) of as little as 17% of their land by 1963. The land reform here did not affect the entire landlord class, but only a very narrow group of rich landlords (having over 200 hectares of land), numbering only 6,000 land owners (holding 15% of all the land, they lost about a third of their property, moreover, most of it — over 70% of the area taken from them — was unarable land).[132]

A very important step was carried out in India in the 1950's: the privileged land ownership (zamindari areas) was liquidated. The next and most important step of the land reforms—the introduction of a ceiling on land ownership—encountered such furious resistance on the part of rich landowners that it brought about practically no changes in the structure of land distribution, that crystallized after reforming the zamindari system. The surplus land taken from the big landowners on the basis of the land ownership ceiling laws and distributed (for payment) among those who tilled the land, by the end of 1963 amounted to only 2.2% of the big (according to Indian standards — 20 hectares and more) land property.[133] That is why the concentration of big land property remains quite high. If in 1953-1954 one per cent of the rural households in India held 17% of the privately owned land, and 5% of the households owned 41% of this land, in 1959-1960 one per cent of the households held 16%, and 5% of the households, 40% of the privately owned land. On the opposite pole in rural areas, after all the reforms the huge mass of landless and semi-landless people have been concentrated. In 1959-1960 70% of the rural households in India owned less than 14% of the land, while 40% of the households owned less than 0.7% of the land.[134]

---

[132] "Report of the Land Reforms Commission for West Pakistan, January 1959", Lahore, 1959, App. 1; "The Second Five-Year Plan. June 1960", Karachi, 1960, p. 185; "Economic Survey of Pakistan 1962-63", Rawalpindi, 1963, p. 31.

[133] Determined from the statistics of: "The National Sample Survey. No. 36. Eighth Round: July 1954-June 1955. Report on Land Holdings (3)", Delhi, 1961, p. 19; "The Third Plan Mid-Term Appraisal. November 1963", Delhi, 1963, pp. 100-101.

[134] "Report of the Committee on Distribution of Income and Levels of Living", pt. I, Delhi, 1964, p. 82.

Tenant farming, chiefly share-cropping, under which the producer, as a rule, pays a rent up to half of his crop and more, remains the scourge of all agricultural development. At the beginning of the 1960's more than one-third of all the arable land was tenant-cultivated in India,[135] and about 50% in West Pakistan. In the Philippines 50.7% of all farms are tenant-cultivated and in Central Luzon, the main rice producing area, the figure is as high as 76%.[136] It is this type of farm that is the first to be deprived of the opportunity to make productive accumulation in agriculture, and to help solve the food problem.

The threat of famine in many emerging nations stems directly from such a conservative agrarian policy, providing for the lengthy preservation of various pre-capitalistic forms of exploitation and the removal of the surplus product from the sphere of agriculture. Life itself, therefore, dictates the necessity of accelerating agrarian transformations and infringing on the selfish interests of a small class group for the sake of saving the nation.

The alternative way is the development of cooperative forms of large-scale farming, the road to which is being paved by profound agrarian reforms. The specific situation existing in the countryside of many developing countries, i. e. the huge "surplus" population, the further parcelization of peasant land ownership and operational holdings, etc., make the task of gradually transforming the low productive economic structures in agriculture along cooperative lines all the more pressing. An example of the UAR has been already cited, where, due to agrarian reforms and the processes of parcelization, the share of petty landownership (up to 2 hectares) increased to 57.1% of the total by 1965. On an average, a petty landowner has now 0.5 hectares of land! It is in the light of this situation that we should assess the importance of the recent trends in the agrarian policy of the United Arab Republic towards a gradual development of peasant cooperative farming.

---

[135] «Проблемы экономического и социального развития независимой Индии», М., 1967, стр. 176 ("Problems of Socio-Economic Development of Independent India", Moscow, 1967).

[136] "International Labour Conference, 51st Session. Report VII (1). Improvement of Conditions of Life and Work of Tenants, Share-Croppers and Similar Caterogies of Agricultural Workers", ILO, Geneva, 1966, p. 6.

The problem of setting up cooperative forms of production has been extensively discussed in recent years in Burma. If we take into consideration a number of measures already taken by the state to reorganize agrarian relations (ordinances to prevent the peasants from becoming landless; law abolishing land rents levied by private owners; gradual removal of trade intermediaries and ousting of money-lenders, etc.), as well as the establishing of a state monopoly on foreign trade of rice—the main farm crop in Burma — it is possible to say that state controls a very powerful lever for promoting a system of big farms on a cooperative basis viewed in historical perspective.

The democratic circles in a country like India uphold the need to pursue a policy of developing cooperative forms of production as a means of boosting the productive forces of agriculture. In this respect, special attention should be given to the recommendations tabled by the Committee of Direction on Cooperative Farming (its Chairman was Prof. D. R. Gadgil). The Committee proceeds from the basic principle that "cooperative joint farming by bringing together, under one management, lands held by a large number of uneconomic holders creates a land base on which technically efficient and economic operations become possible... Moreover, as the capital resources of the uneconomic holders are notoriously meagre, these have to be appropriately enlarged in the initial stages to bring about needed technical improvements in the management of pooled lands". In implementing such measures, the productivity of labour will naturally increase. As the Committee stresses, "efficient management of pooled lands implies that the total input of labour on the cooperative joint farms, for the same types of productive enterprises, is less than that on scattered individual holdings of which they are composed". However, at the same time, if the development proceeds in the direction of intensifying agricultural production, "there would be some permanent increase in the level of demand for labour on the cooperative joint farm", even though, it is quite understandable, the possibility of increasing employment on this basis is "somewhat limited".[137] In the opinion of the Committee, "the class of small or 'uneconomic' cultivators, whose number is 'very large' account

---

[137] "Report of the Committee of Direction on Cooperative Farming", Delhi, 1965, p. 183.

for no less than 15% of all the cultivated land. And it is precisely this layer of farmers who should form cooperatives first of all. At the same time "categories of farmers, other than small holders, may also form joint cooperative farming societies and these may achieve a much better utilisation of pooled resources through joint efforts". In India, however, the problem lies in that "the national programme of cooperative farming is not directed essentially towards these...."[138]

In certain countries (for instance, in the UAR), broad opportunities for developing large-scale highly commercialized production on the basis of state or cooperative farms are offered by the policy of huge irrigational systems, permitting vast tracts of state land to be cultivated and put to good use. This policy is of vital social significance. The successful implementation of this policy considerably broadens the material basis for profound democratic transformations in agrarian economy as a whole. In the sphere of agricultural production there is now a really powerful alternative for taking the place of the capitalist structure, which in many countries has now centralized a significant part of production, and what is more important, marketed products. In the prevailing conditions if the state does not possess a certain minimum of material resources (for instance, different technical means used for helping peasant economy), and without establishing definite minimum prerequisites for developing large-scale production of a public type, immediate restrictions on production in the capitalist structure, in several cases, could retard the overall growth of agricultural production, and especially the output of marketed products, with the result that the food crisis will be aggravated.

Agrarian transformations in themselves, even if they are the most efficient from the social point of view would have only a minimum influence on production, unless conditions were provided at the same time for economic stimulants coming into play in full force in agriculture. That is why it is no less important for a national state striving to pull up agricultural production by its bootstraps, to work out and implement a policy based on the demands of the law of value. Such a policy has several aspects. One of them is that the state steps resolutely into the sphere of price formation on the national markets in order to considerably improve

[138] *Ibid.*, pp. 183, 184.

the system of prices which the cultivators receive for their produce. This policy is all the more imperative because of the further development of commodity and money relations in the national economy, including rural areas, on the one hand, and in view of the growing non-equivalent exchange between the industrial sector and the ever-lagging agriculture, on the other. On a national plane, the effective implementation by the state of a price policy that is in accord with the demands of the law of value presupposes that trading-cum-speculative capital is squeezed out of the market (highly restricted in the initial stages), while the institutional vacuum is filled by state and cooperative marketing organizations. It is, therefore, not at all surprising that in many developing nations different social layers and groups are engaged in a heated discussion and struggle over the problem of agricultural prices, the scale and forms whereby to regulate them, since any changes in this sphere will directly affect diametrically opposite class interests.[139] On an international plane, this policy is called upon to protect national food production from the disastrous influence of the laws of the world capitalist economy.

The raising and stabilization of prices received by the agricultural producers are not the only means for ensuring economic incentives to boost agricultural output. In this respect, broad opportunities are opened up also by the policy of the state subsidizing (i. e. the state covers part of the selling price) the means of production purchased by the producer.

Due to the imperialist monopoly rule on the world market wherefrom the developing nations continue to obtain a significant or greater part of modern machinery and other means of production for their agriculture, or industrial equipment necessary for their production, the ultimate prices which these means of production are sold for on the home market of these countries and introduced in the production process are exceedingly high. As a rule, they are higher than the corresponding prices in the developed capitalist countries. For instance, in 1964/65 farmers in France had to pay $25.5 for 100 kg of ammonium sulphate (in terms of 100%

---

[139] For details on these questions see: Е. И. Миронова, Продовольственная проблема в Индии (зерновой рынок и государственная политика), М., 1967 [E. I. Mironova, *The Food Problems in India* (*The Grain Market and the State Policy*), Moscow, 1967].

nutritional content), the most commonly used chemical fertilizer, in the Federal Republic of Germany—26.2, in the U.S.A.—27.8, while cultivators in the UAR had to pay $31.3 and in the Philippines — 35.6; in India in 1963/64 they bought this fertilizer for $36.7 per 100 kg of nutritional content equivalent.[140] It is common knowledge that in the developed countries the average fertilizer-purchasing farm is much larger and more productive than in the developing countries. Naturally, in view of the high prices, the potential demand for this very important means of production in the second group of countries is highly limited (nothing is changed by the fact that in certain national states, as in India for instance, there is an acute shortage of chemical fertilizers, as a result of which even with the extremely high prices it is impossible to satisfy the existing narrow demand for them).

The policy of subsidizing the selling price of modern means of production is aimed precisely at greatly expanding the contingent of farmers utilizing these means of production. This policy is practised in certain developed capitalist countries (Great Britain, for instance). It is resorted to by the governments in certain developing countries, though primarily in respect of a very limited scope of means of production, mainly chemical fertilizers (for example, in Pakistan).[141]

With a general increase in market prices, the stabilization of prices on this or that type of means of production should be regarded as one of the variants of the policy of subsidizing their selling prices. For instance, in the UAR in 1965 the wholesale price index for chemical fertilizers was only 99 (1952 = 100), while the wholesale price index for grain went up to 154 points, and for the group of industrial goods (finished articles and raw materials) the index grew to 124 points.[142] This was one of the reasons that made it possible for such an extensive utilization of these vital means of production in the agriculture of the UAR.

In certain circumstances (for example, if a country, for one reason or another, has difficulty in simultaneously

---

140 "Production Yearbook 1965", FAO, Rome, 1966, pp. 404-422.

141 In Gt. Britain in the first half of the 60's, government subsidies for nitrogen and phosphate fertilizers amounted to 36-42%. In Pakistan they amounted to 53%; however, the bulk of the fertilizer was used by a very narrow group of farms, mainly belonging to rich landowners.

142 "National Bank of Egypt Economic Bulletin", Cairo, vol. XIX, 1966, No. 4.

carrying out a whole complex of measures to economically stimulate agricultural production through the mechanism of the market), the policy of subsidizing the selling prices on the means of production utilized by the farmers may even be more preferable than the policy of raising the general level of prices on agricultural products. As was noted earlier, part of the new, increased price on farm produce will inevitably enter the consumption fund of the producer (especially when the case concerns the petty producer), i. e., it is spent in the non-productive sphere. And on the contrary, the policy of selling the means of production (including water from state irrigational systems) at prices lower than the market prices, which sharply cuts the cost of production, averts such a possibility. And even though, in this case, the state will have to cover the expenditures directly (they can be regarded as one of the forms of state investments), in the long run, they will be offset by the growing volume of agricultural produce.

The urgent need to do away with agricultural backwardness puts in the forefront the problem of establishing new proportions in redistributing the national fund of capital investments between different sectors of the economy. In fact, it is a matter of essentially increasing investments in agriculture, including those through the public sector. Some changes in this direction are already in the making. The grave food difficulties, as stated by ECAFE, "gave greater urgency to the rethinking of member Governments in the region in centering their programmes towards expanding food output, particularly foodgrains".[143]

The significance of branch-wise distribution of capital investments for raising agricultural output is vividly reflected in the example of Thailand and the Philippines — two countries with approximately the same socio-economic system (though, perhaps, in the Philippines, the semi-feudal structure is stronger than in Thailand), general level of development and comparable specializations of agriculture (grain production, mainly rice, as the chief branch). In the period from 1960 to 1965 capital investments in agriculture in Thailand amounted to 15-20% of the gross fixed capital formation in the economy as a whole, while in the Philippines this figure varied from 3.8 to 7.5%. Industry in Thailand received a share of 22-24% of the gross fixed capital formation, while

---

[143] "Economic Survey of Asia and the Far East 1966", p. 108.

in the Philippines the corresponding figure was 21-29%. Whereas in the Philippines the agricultural sector occupied an insignificant place in the national programmes of development, in Thailand much more attention was given to its consolidation. And this was by far not the least factor that made Thailand's gross agricultural product grow much faster than in the Philippines (for example, the average annual growth rate of agricultural production in Thailand in 1960-1965 amounted to 4.6%, as compared to 3.2% for the Philippines).[144]

The most formidable objective difficulty hindering technical re-equipment of argiculture in the developing nations lies in their weak industrial potential. Moreover, it appears that the least developed sector of industry is that which produces means of production for agriculture. The serious obstacle is also the weekness of the system of disseminating knowledge about modern methods of farming among the farmers. The removal of these "bottlenecks" may, undoubtedly, require considerable additonal capital investments.

Although there can be no doubt, in certain cases, as to the need for changing the structure of national capital investments (especially, investments by the state), nevertheless, in a class society with vividly pronounced class polarization, as is the case in many developing countries, its real solution is extremely complicated because such a change can adversely affect the interests of different social groups. And this, in turn, cannot but lead to bitter clashes in the sphere of politics. Such an example can be seen in the struggle which has developed in India in recent years in respect to the goals and tasks of the Fourth Five-Year Plan, the fulfillment of which was put off for 4 years (1969/70). This is likely to hamper the entire process of creating the material potential for resolving the food problem.

It goes without saying that the allocation of national investments in conformity with the demand for the proportional development of all sectors of the economy can yield all the greater an effect, the deeper the social changes which create a primary prerequisite for the most efficient use of investments, assigned for both agriculture and the industries catering to it.[145]

---

144 *Ibid.*, pp. 104, 105.

145 Proceeding from a clear understanding of this situation the Planning Commission of Government of India, for example, emphasizes:

Meanwhile, the developing countries are in dire need of agricultural capital investments. According to the calculations made by a group of American experts, in order for the developing countries of the world to double agricultural production, the consumption of chemical fertilizers (in terms of nutritional content) must be increased to 67 million tons, against 6 million in 1966. To produce such an amount of fertilizer and to distribute it among the farms in the period up to 1985 would require an investment of $17 billion. In order to make use of this mass of fertilizer, it would be necessary to make huge capital investments in irrigation. Besides that, additional capital investments would be necessary to disseminate among the cultivators agrotechnical know-how, to combat pests, and so on.[146] All this runs into fabulous sums. For instance, the investments for producing the needed amountof chemical fertilizer total 8% of the gross domestic product in all the emerging nations of the world in 1965. But on the other hand, this sum does not even amount to 1.2% of the gross domestic product produced in 1965 by all the economically developed countries (capitalist countries and socialist lands—U.S.S.R and Eastern Europe) and, in particular, this figure is approximately half of the U. S. current expenditures on the war in Vietnam ($30 billion a year). Tremendous national incomes could be channelled towards productive use by carrying out general disarmament—herein lies one of the most important sources for capital investments to combat the threat of world famine.

The food problem is a social problem whose roots go down into the structure of the socio-economic system. At the same time, it is a major agrotechnical problem. Its future solution depends on systematically carrying out a complex policy, all the elements of which are closely inter-twined. And in this modern age the fate of hundreds of millions of people depends on how this policy is put into effect.

„Delays in implementation (of land reforms.— *V. R.*) are apt to... hamper execution of production programmes... Unless the reforms are speedily implemented the benefits of the proposed large-scale investment in the Fourth Plan in agriculture in the public sector will not accrue to the rural poor..." ("Implementation of Land Reforms. A Review by the Land Reforms Implementation Committee of the National Development Council", New Delhi, 1966, p. 18).

[146] "The World Food Problem. A Report of the President's Science Advisory Committee", vol. I, p. 86.

World Food Supply

An Arno Press Collection

Agricultural Production Team. **Report on India's Food Crisis & Steps to Meet It.** 1959

Agricultural Tribunal of Investigation. **Final Report.** Presented to Parliament by Command of His Majesty. 1924

Bennett, M. K. **The World's Food:** A Study of the Interrelations of World Populations, National Diets and Food Potentials. 1954

Bhattacharjee, J. P., editor. **Studies in Indian Agricultural Economics.** 1958

Brown, Lester R. **Increasing World Food Output:** Problems and Prospects. 1965

Brown, Lester R. **Man, Land & Food:** Looking Ahead at World Food Needs. 1963

Christensen, Raymond P. **Efficient Use of Food Resources in the United States.** Revised Edition. 1948

Crookes, William. **The Wheat Problem.** Revised Edition. 1900

Developments in American Farming. 1976

Dodd, George. **The Food of London.** 1856

Economics and Sociology Department, Iowa State College. **Wartime Farm and Food Policy,** Pamphlets 1-11. 1943/44/45

Edwards, Everett E., compiler and editor. **Jefferson and Agriculture:** A Sourcebook. 1943

Famine in India. 1976

Gray, L. C., et al. **Farm Ownership and Tenancy.** 1924

Hardin, Charles M. **Freedom in Agricultural Education.** 1955

High-Yielding Varieties of Grain. 1976

[India] Famine Inquiry Commission. **Report on Bengal.** 1945

Johnson, D. Gale. **Forward Prices for Agriculture.** With a New Introduction. 1947

King, Clyde L., editor. **The World's Food.** 1917

Marston, R[obert] B[right]. **War, Famine and our Food Supply.** 1897

Mosher, Arthur T. **Technical Co-operation in Latin-American Agriculture.** 1957

The Organization of Trade in Food Products: Three Early Food and Agriculture Organization Proposals. 1976

Projections of United States Agricultural Production and Demand. 1976

Rastyannikov, V. G. **Food For Developing Countries in Asia and North Africa:** A Socio-Economic Approach. Translated by George S. Watts. 1969

Reid, Margaret G. **Food For People.** 1943

Schultz, Theodore W., editor. **Food For the World.** 1945

Schultz, Theodore W. **Transforming Traditional Agriculture.** 1964

Three World Surveys by the Food and Agriculture Organization of the United Nations. 1976

U. S. Department of Agriculture, Agricultural Adjustment Administration. **Agricultural Adjustment:** A Report of Administration of the Agricultural Adjustment Act, May 1933 To February 1934. 1934

U. S. Department of Agriculture. **Yearbook of Agriculture, 1939:** Food and Life; Part 1: Human Nutrition. 1939

U. S. Department of Agriculture. **Yearbook of Agriculture, 1940:** Farmers in A Changing World. 1940

[U. S.] House of Representatives, Committee on Agriculture. **Oleomargarine.** 1949

[U. S.] National Resources Board. **Report of the Land Planning Committee. Part II.** 1934